99% APE

HOW EVOLUTION ADDS UP

EDITED BY

JONATHAN SILVERTOWN

AUTHORS

CAROLINE M. POND, JONATHAN SILVERTOWN,
DAVID ROBINSON, PETER SKELTON, DANIEL NETTLE,
MONICA GRADY AND GARY SLAPPER

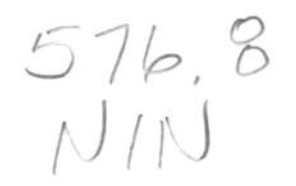

THE UNIVERSITY OF CHICAGO PRESS
CHICAGO AND LONDON

Acknowledgements

The authors are grateful to the following for their comments on the manuscript:
Kim Bryan, Rissa de la Paz, Kat Hull, Michael Majerus, James Moore and Michael Reiss.

Jonathan Silvertown is professor of ecology in the department of biological sciences at the Open University. He is the author of *Demons of Eden* and *An Orchard Invisible*, both published by the University of Chicago Press.

The University of Chicago Press, Chicago 60637
The University of Chicago Press, Ltd., London

Printed in China

18 17 16 15 14 13 12 11 10 09   1 2 3 4 5

ISBN-13: 978-0-226-75778-0 (paper)
ISBN-10: 0-226-75778-1 (paper)

First published in Great Britain in 2008 by the Natural History Museum, Cromwell Road, London SW7 5BD, in association with The Open University, Walton Hall, Milton Keynes MK7 6AA.

Library of Congress Cataloging-in-Publication Data

99% ape : how evolution adds up / Jonathan Silvertown ... [et al.].
p. cm.
Includes bibliographical references and index.
ISBN-13: 978-0-226-75778-0 (pbk. : alk. paper)
ISBN-10: 0-226-75778-1 (pbk. : alk. paper) 1. Evolution (Biology)—Popular works. I. Silvertown, Jonathan W. II. Title: Ninety-nine percent ape.
QH367.A13 2009
576.8—dc22
2008043807

∞ The paper used in this publication meets the minimum requirements of the American National Standard for Information Sciences—Permanence of Paper for Printed Library Materials, ANSI Z39.48-1992.

Contents

Introduction

DARWIN WAS MOCKED FOR SUGGESTING THAT humans have apes for ancestors, but every scientific advance in the study of life in the last 150 years has confirmed the reality of evolution. The aim of this book is to present the latest evidence for the general reader. You don't need any background in science to enjoy this book. After all, it is about something that everyone should know. We share about 99% of our DNA with chimps and this common ancestry has the deepest implications for how we see ourselves. The greatest question, *How does evolution happen?* was solved by Charles Darwin a century and a half ago. Even today Darwin's brilliant idea, natural selection, is the only mechanism we know of that can produce adaptation.

In this book of four parts, we begin with **Origins**. The fundamental similarities of all living things point to a single origin of today's life on Earth, which means we are kith and kin to crabs and cacti. The tree of life is no longer a metaphor, but a genealogy of all living things that even now is being built from clues to ancestry hidden in the genetic code of every living thing. Darwin guessed that the human species had evolved in Africa, and the fossil evidence now shows that he was right about that too. He also wrote that it was difficult to imagine how structures as complicated and sophisticated as eyes could have evolved by gradual steps through natural selection, but in **Body building** we discover that it was easier then anyone imagined. Just as revelatory is the unfolding story, illustrated by recent fossil finds and the study of gene action in surviving species of ancient lineages, of how our fish ancestors first crawled out onto land. No less amazing are the fossils and genetic studies that chronicle how whales and dolphins evolved from terrestrial ancestors. And if you thought that all the dinosaurs were extinct, we have a surprise for you.

We can see in the genes where **Diversity** comes from, how sex shuffles the deck and deals each new generation a novel and yet familiar hand. The raw material of genetic variation is winnowed to produce new species from old and can produce a lakefull of unique fishes, an archipelago of Darwin's finches or the evolutionary radiation of the flowering plants and the huge diversity of insects that pollinate and feed upon them. For our self-obsessed species, human diversity is just as fascinating and turns out to be very recent. We are all Africans, and since the bulk of the human population no longer lives in Africa, we are nearly all descended from immigrants. And so to the **Here and now**. Evolution is not just relevant to the past, but it is also about the present and the future. It matters now because emerging diseases like HIV and pandemic flu are evolutionary opportunists, crossing the species barrier from other primates (HIV) or our domestic livestock (flu) when they acquire the genes to do so, either from other microbes

LEFT These footprints, made by two hominids in Africa some 3.5 million years ago, attest to a very recent episode in the long history of life on earth.

or by mutation. In the war against such diseases, evolutionary biology is in the front line of our defence.

If one thing sets us apart from our primate relatives, it is our minds. What can evolutionary psychology tell us about why we behave as we do? Religion has a logical problem with evil (why does God permit it?), but Darwinism has a problem with good, or so it might appear. How can natural selection, which favours individual success, permit the evolution of co-operation and other acts that we call 'good'? Is there an evolutionary explanation for morality? If science can illuminate the source of morality, once the exclusive realm of religion, is there any place left for God? Is a belief in God compatible with the science of evolution, or must all believers become creationists and all evolutionists adopt atheism? From the burning issues of the present, inevitably to the future: What next? Are we still evolving? What does the future hold, not just for us, but for the other species with which we share our evolutionary history and this planet?

CHAPTER 1 99% Ape

ON 24 JULY 2007, FIVE BULGARIAN NURSES and a Palestinian doctor boarded a plane in Tripoli, Libya. This was no ordinary flight. The plane belonged to the President of France and also on board was the wife of the newly elected French President, who had travelled to Libya to escort the six medical workers out of the country. The six had been in Libyan prisons for eight years, accused of deliberately infecting more than 400 children in the hospital where they worked with the Human Immunodeficiency Virus (HIV) that causes AIDS. During their incarceration they were tortured and twice sentenced to death by firing squad. Finally, in a political deal with the European Union, which agreed to pay for the treatment of those children who were still alive, the medical workers had their sentences commuted to life imprisonment and they were allowed to leave Libya.

BELOW Six medical workers in the dock of a Libyan court where they stood acccused of infecting child patients with HIV.

During their long ordeal, the plight of the six medical workers had become a *cause célèbre*, with 114 Nobel Prize-winners petitioning on their behalf. Luc Montagnier, the discoverer of HIV, testified that the children's HIV infections were acquired in the hospital from poor hygiene and that the medical workers were innocent, but his testimony was ignored by the Libyan court. Despite this, one piece of scientific investigation appears to have been decisive in winning the arguments that took place between EU, British and Libyan diplomats behind closed doors. In any negotiation it helps to be certain of your facts. The crucial piece of scientific evidence that exonerated the medical workers came from a state-of-the-art study of how HIV had evolved.

Evolution liberates

Evolution means change through successive generations, and the AIDS virus evolves very quickly indeed. It is doubly difficult to produce an effective vaccine against HIV because it presents an ever-moving target and because the virus attacks the immune system itself. Viruses are tiny particles that depend entirely upon the cells of their host to reproduce.

OPPOSITE Darwin held a mirror up to the human species and saw the features of an ape.

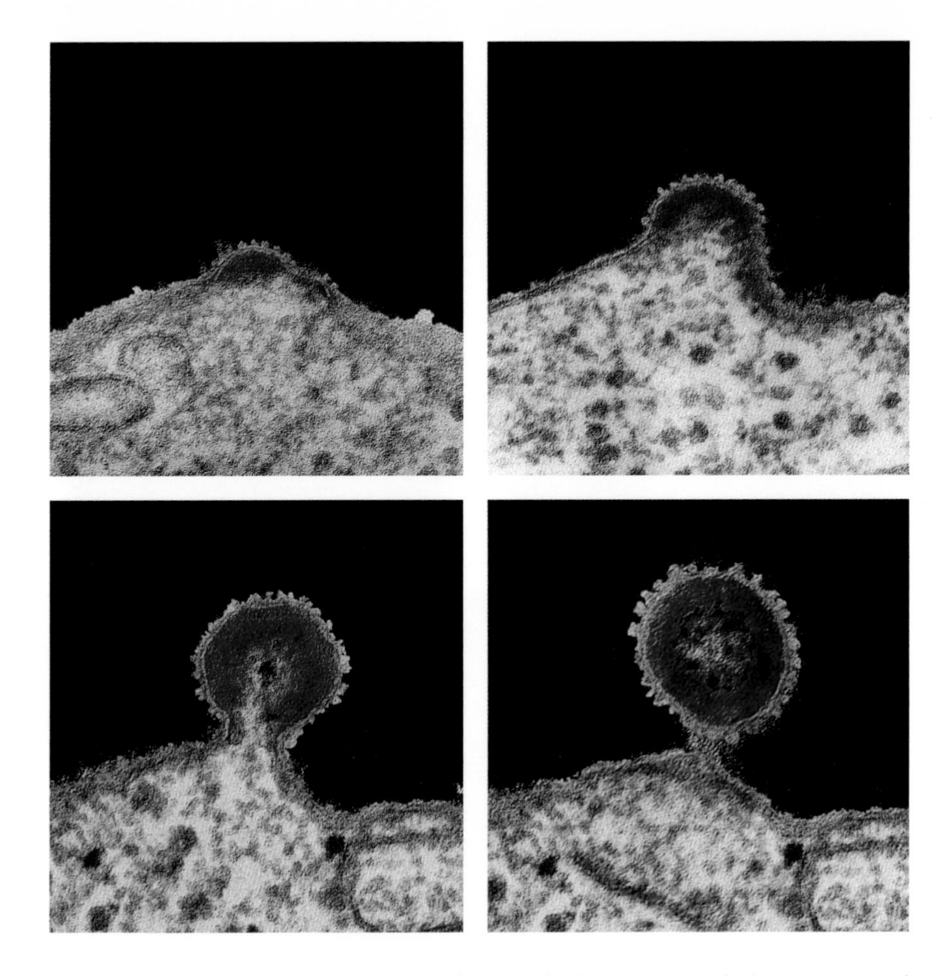

RIGHT Images of the Human Immunodeficiency Virus (HIV) emerging from a cell.

They are the ultimate parasites, stripped down to the barest essentials that are required to infect cells and then hijack the cellular machinery inside to produce more copies of themselves. Those bare essentials are just the genetic blueprint for the virus itself, plus a coat that protects it and helps insert the blueprint into the cells of its victims. HIV has only nine genes of its own, but it uses these to hijack over 200 genes belonging to the cells of its host.

The genetic blueprint of HIV is held in a molecule called RNA (which is like a single-stranded version of DNA) that host cells can read, rather as you are reading the words on this page. Imagine that you now had to copy the text in this book by re-typing it. Even if you are a good typist, the new version would be certain to contain some errors. Copy the copy and there would be some more. This happens when viruses replicate too. The changes that take place are called mutations.

Mutations generally occur at a very low rate. Each of us on average has in a lifetime only about 1.6 new ones among our 20,000–25,000 genes. This is because our cells use error-correcting machinery to fix mistakes as the blueprint is replicated (a little like the computer software that spell-checks as you type). HIV doesn't use machinery like this, so it makes a lot of errors. The changes are an example of evolution. Most of the genetic mutations that occur in HIV are probably damaging to the viruses that carry them, but there are so many millions of virus particles that some are almost bound to carry changes that help the virus escape the body's defences or resist anti-AIDS drugs. Any virus particles carrying such beneficial (to them) mutations multiply and spread faster than those without. They are favoured by natural selection. If natural selection had a motto, it might be 'Nothing succeeds like success'.

HIV evolves so fast that individual patients who have been infected for more than a short time have many genetically different HIV particles in their bloodstream. Different patients likewise carry their own set of HIV variants. By reading the code of a RNA molecule, called its sequence, we can compare it with others and use the differences to identify where mutations have occurred. This information can then be used to work out an evolutionary tree, which traces the genealogical relationships among virus particles using the same principle by which a teacher can detect who has copied whom in a class where several pieces of homework seem suspiciously alike. Right answers may look alike because the scope for variation among correct answers is limited, but there is a multitude of ways to get something wrong, so repeated mistakes of the same kind are tell-tales. Any particular error should be rare and if it shows up improbably often, copying may have taken place. By the same token, rare mutations (errors) that occur in the same place in the sequence of two viruses mean that both may have been copied from the same ancestral sequence. When there are sufficient mutations, the evidence for 'who copied whom' can be quite conclusive.

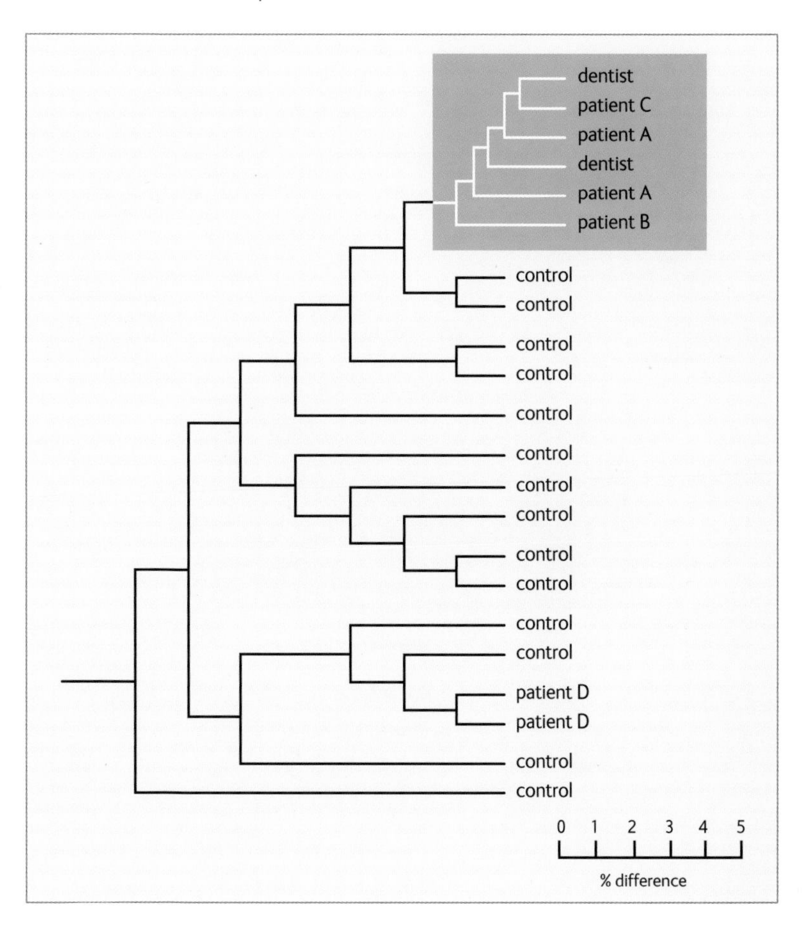

ABOVE Evolutionary tree for HIV infections in a dentist and others with the virus, including controls and four of the dentist's patients. Viruses from patients A, B and C are very similar to those carried by the dentist. Patient D carries a virus apparently acquired from another source .

An evolutionary tree for any set of viruses (or other evolving organisms) is analogous to a genealogy for a family of people. The valuable thing about an evolutionary tree when dealing with a virus such as HIV is that it can be used to trace the source of an infection. The first time this was done with the AIDS virus was when an evolutionary tree was constructed for viruses from an HIV-positive dentist and his patients, whom he had been accused of infecting during dental practice. The tree showed that the HIV virus in three of the patients (A, B and C in the figure above) had indeed all evolved from those carried by the dentist, while a fourth patient (D in the figure above) had contracted HIV from another source.

If you know the rate of mutation, it is possible not only to reconstruct an evolutionary tree for the viruses carried by different HIV patients, but also to estimate when the viruses last shared a common ancestor. This is like meeting a distant relative and working out

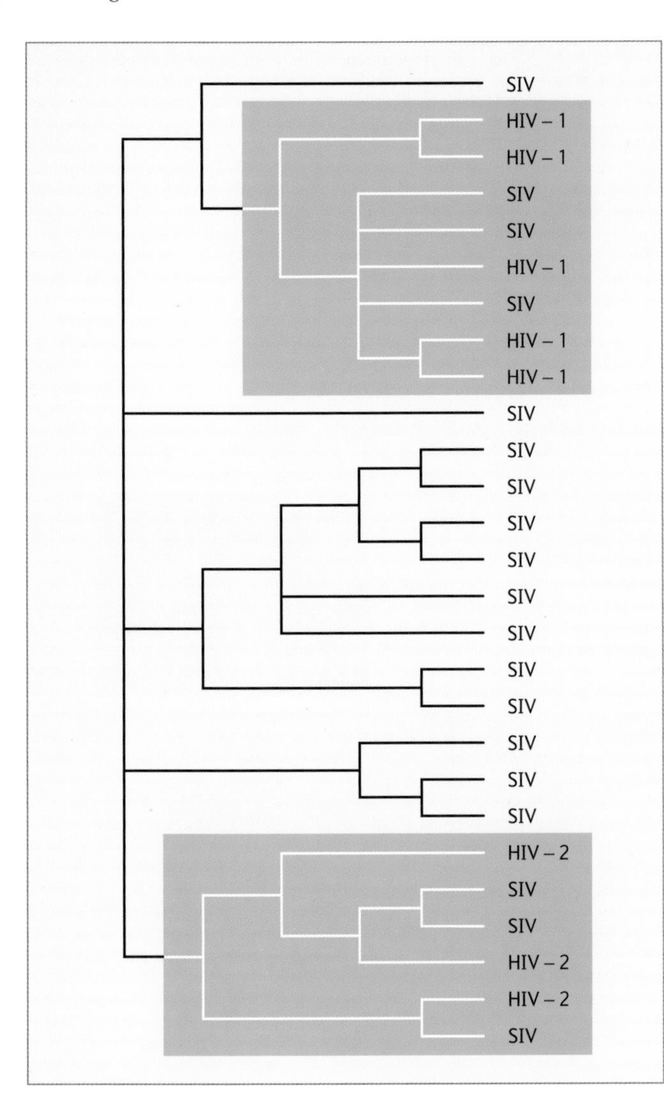

ABOVE An evolutionary tree showing the relationships between two strains of the Human Immunodeficiency Virus (HIV-1 and HIV-2) and the Simian Immunodeficiency Virus (SIV) found in other primates. HIV-1 evolved from chimp SIV and HIV-2 from the SIV that infects sooty mangabeys.

how many generations you need to go back to find the connection between you. If you are first cousins, you share grandparents; if you are second cousins, you share great-grandparents, and so on.

In the summer of 2006, as the Libyan court was in recess considering its verdict in the appeal of the six medical workers, scientists at the Universities of Oxford and Rome worked against the clock to compile an evolutionary tree for the different virus strains that had been isolated from children in the case. When they were published in December 2006, the evolutionary tree powerfully supported the case that there had been a terrible miscarriage of justice. It demonstrated that the viruses that infected the children had been spreading among patients in the hospital, including some of the children, before the medical workers had even arrived there in March 1998. There was no way that they could plausibly be blamed for the infections, either directly or indirectly.

Evolutionary trees

We shall look more deeply into the evolution of disease-causing organisms in Chapter 15 *Catch me if you can*, but the example of HIV illuminates another evolutionary question too. Where did HIV come from? When cases of AIDS first came to medical attention in the late 1970s the causative agent was unknown. Even after HIV had been discovered and it had been established that the cause of AIDS was a hitherto unknown virus, its origins were a mystery. Conspiracy theories about the virus having been manufactured, escaped or been deliberately spread were seriously entertained, even by some eminent scientists. But, the evolutionary tree of the virus eventually told a different story.

HIV evolved from a very similar virus called the Simian Immunodeficiency Virus (SIV) that infects our close primate relatives. The evolutionary tree shows how closely related HIV and SIV are. There are two forms of HIV, HIV-1 and HIV-2, and the tree shows that they evolved independently. HIV-1 evolved from SIV in chimps and HIV-2 evolved from SIV in sooty mangabeys. It should have been no surprise that a rapidly mutating virus affecting apes and monkeys could evolve into a form that can infect humans. We are, after all, apes ourselves.

Darwin, apes and Victorians

How "frightful and painfully and disagreeably human!" exclaimed Queen Victoria when she encountered one of the first orang-utans to be exhibited at London Zoo in 1842. The Queen's impression would have been reinforced by the Victorians' predilection to clothe their primate exhibits. Perhaps the clothing itself was a subconscious recognition of the human affinities of apes. Queen Victoria and most of her subjects believed in the biblical account of human origins. According to the Book of Genesis, humans were created in God's likeness and all species of animals and plants were each separately created. Each type was fixed and similarities, like the resemblance of orang-utans to humans, were just proof that God created life using common plans. The Victorian naturalist Charles Darwin, however, thought otherwise.

By 1842, Darwin had come to the conclusion that species were not fixed and that the similarities among primates, for example, were a consequence of evolution from a common ancestor. He believed that the resemblance between humans and orang-utans was evidence of

BELOW An orang-utan. Orang-utan means 'man of the forest' in the Malay language. We share a common ancestor with these and other great apes.

RIGHT Jenny the orang-utan at London Zoo in 1842. Queen Victoria declared the ape "frightful and painfully and disagreeably human".

Jenny, the orang-outang, taking tea at the Zoo.

BELOW Darwin the ape-man, as ridiculed by a Victorian cartoonist.

'common descent', or evolution, and he had worked out a theory to explain how it took place. This notion was so heretical in early Victorian England that Darwin dared not publish his theory until he had collected a mountain of evidence to support it. Even 17 years later in 1859 – when he was forced into publishing *On the Origin of Species* by the fear that another naturalist, Alfred Russel Wallace, who was developing a similar theory, would gain precedence – Darwin regarded the book as only a sketch of the more complete case that he wanted to make in support of his theory. From the very beginning, then, the case for Darwinian evolution has rested upon evidence, not just belief. One hundred and fifty years on from the publication of *On the Origin of Species*, the evidence for evolution is overwhelmingly strong. This book, *99% Ape*, aims to show you how that evidence adds up.

Charles Darwin did not invent the idea of evolution. His many predecessors who believed that species change included his own grandfather, Erasmus Darwin (1731–1802), a physician who liked to attend his patients, in the vicinity of Lichfield in the English Midlands, in a carriage blazoned with a family crest that bore the motto *E conchis omnia* – 'Everything from seashells'. We now know that Erasmus was right – life did begin in the sea – but this was an eccentric, not to say scandalous, opinion in the 18th century and Charles did not have the stomach for ridicule that his grandfather evidently did. Charles' father Robert, also a doctor, held heretical views too, but kept his own counsel. Small wonder, then, that Charles was cautious. When he first tentatively broached the subject of evolution with his friend Joseph Dalton Hooker in a letter written in January 1844, he wrote:

> *... I was so struck with distribution of Galapagos organisms &c &c and with the character of the American fossil mammifers [mammals], &c &c that I determined to collect blindly every sort of fact, which could bear any way on what are species. — I have read heaps of agricultural and horticultural books, and have never ceased collecting facts — At last gleams of light have come, and I am almost convinced (quite contrary to opinion I started with) that species are not (it is like confessing a murder) immutable.*

BELOW Erasmus Darwin (1731-1802), Charles's grandfather, was an evolutionist who expounded his theories in verse.

What made Charles Darwin different from all earlier evolutionists was his discovery of how evolution works. He discovered how evolution produces changes that make organisms better adapted to their environment, better able to defend or feed themselves, and different from their ancestors. The name Darwin gave to his big discovery – "my theory", he called it – was natural selection.

Viruses were unknown in Darwin's day, but natural selection explains how they evolve. This remarkable achievement is testament to the enduring power of Darwin's discovery – a theory so far-reaching that 150 years later we are still, daily, finding fresh examples of natural selection at work.

In the closing pages of *On the Origin of Species*, Darwin commented that:

> *Authors of the highest eminence seem fully satisfied with the view that each species has been independently created.*

But he foresaw the revolution in understanding that his theory of evolution would produce and *On the Origin of Species* is packed with fertile insights. Many of these provide a starting point for the exploration in this book of the modern relevance of the theory of evolution.

LEFT Charles Darwin's study at Down House in Kent, where he wrote *On the Origin of Species*.

DARWIN'S INSIGHTS

'Light will be thrown on the origin of man and his history.'

Although he lacked any fossil evidence, Darwin supposed that the human species evolved in Africa. Subsequent fossil finds and genetic research have not only shown that Darwin was uncannily right, but also how and when modern humans evolved (Chapter 5 *African genesis*) and spread from Africa (Chapter 14 *The race from Africa*).

'I believe that animals have descended from at most five progenitors, and plants from an equal or lesser number. Analogy would lead me one step further, namely, to the belief that all animals and plants have descended from some one prototype.'

We now know that there is a single tree of life, and that all animals belong to one branch and all plants to another (Chapter 3 *The tree of life*).

'As natural selection acts solely by accumulating slight, successive, favourable variations, it can produce no great or sudden modification; it can act only by very short and slow steps.'

The view that evolution occurs gradually, with no sudden leaps, has been challenged several times by evolutionists seeking alternative or supplementary mechanisms for evolution, but it has successfully stood the test of time (Chapter 2 *Darwin's brilliant idea*).

'The whole history of the world … will hereafter be recognized as a mere fragment of time, compared with the ages which have elapsed since the first creature, the progenitor of innumerable extinct and living descendants, was created.'

Fossil evidence now suggests that life first evolved on Earth at least 3500 million years ago (Chapter 4 *First life*).

'But, as by this theory innumerable transitional forms must have existed, why do we not find them embedded in countless numbers in the crust of the earth? … I believe that the answer mainly lies in the [geological] record being incomparably less perfect than is generally supposed.'

Darwin was right that the fossil record was imperfect, but some of that imperfection has been put right by modern geologists' improved understanding of where to look for fossils. As a result, fossils that were once regarded as 'missing links' have now been discovered for many of the most important transitions in the history of life, including the first animals with backbones to invade the land from the sea (Chapter 7 *A fish out of water*), the evolution of whales from land mammals (Chapter 8 *A whale of a problem*), and the evolution of birds (Chapter 9 *Feathered fossils*).

'Psychology will be based on a new foundation …'

How and whether Darwin was right about this is still a matter of fierce scientific debate, but there can be no doubt that the human mind did evolve. The open question is to what degree the evolutionary history that shaped the brain constrains us to behave in particular ways (Chapter 16 *Darwin in mind*, Chapter 17 *Why be good?*).

What's in 1%?

What does it mean to say that humans and apes, or more specifically chimpanzees, are 99% similar at the genetic level? How can a 1% difference between chimps and humans explain all the obvious differences between the species *Pan troglodytes* (chimpanzees) and *Homo sapiens* (us)?

The answer lies buried in the genome sequences of humans and chimps, and in the roughly 3 billion letters of the genetic code that each of them contains. Laid side-by-side, the immensely long sequence of letters in the recipe for how to build a chimp is *almost* identical to the recipe for a human. The difference is just 1.06%. However, there is more in that inconsequential-sounding fraction than at first appears. Even 1% of 3 billion is 30 million – quite a big number. So, how can 30 million genetic letters spell the difference between chimp and human?

To answer this question we need to know how many of those letters change the spelling of words in the recipe, and of those words that are changed how many actually alter the recipe itself. It turns out that the 1.06% letter differences between chimp and human represent a meaningful change to about 6% of the genes in the primate recipe, and that many of those genes may be switches that turn other genes on or off, with consequently bigger effects than their small numbers might suggest. As science reveals more and more of how genes work, it is becoming easier to fathom how small genetic changes can add up to significant changes in appearance, behaviour and intelligence like those that took place over the short space of about 6 million years since chimps and humans parted company from our common ancestor. How evolution makes the small steps that lead to big change was Charles Darwin's brilliant idea.

BELOW The primate family tree showing representatives of its living members. Dates of divergence are approximate; mya, million years ago.

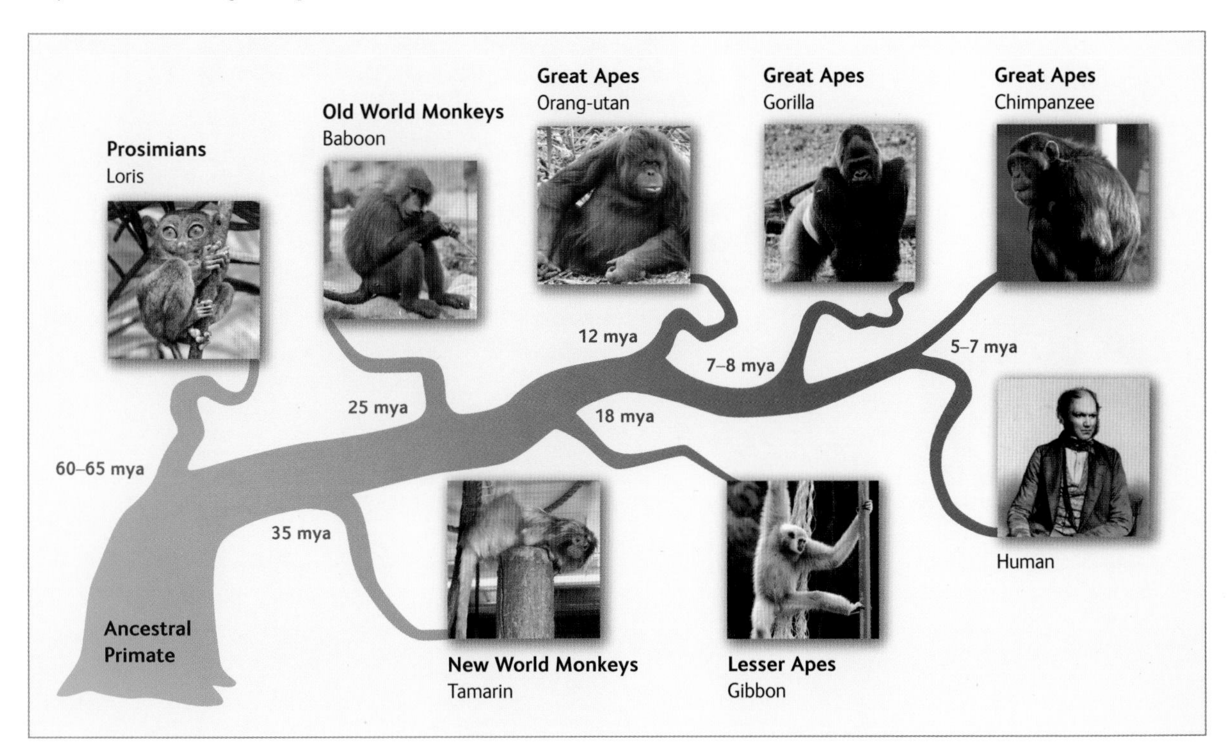

BELOW A pocket watch; the Rev William Paley compared its mechanism to the design of living things.

CHAPTER 2 *Darwin's brilliant idea*

IMAGINE THAT YOU ARE OUT FOR A WALK and that you come across a watch lying upon the ground. You pick it up, inspect it carefully and discover that it contains an intricate and finely crafted mechanism, which appears to have been designed to tell the time. It is self-evident, said the 18th-century cleric William Paley (1743–1805), that since the watch was designed, there must therefore have been a designer. By analogy, the intricate mechanisms of life that are so well designed for animals and plants to function in themselves and with each other are evidence of a designer. That designer must be supremely intelligent and powerful – the designer can only be God.

BELOW Rev William Paley (1743-1805), author of *Natural Theology*, in which he argued that design and apparent purpose in nature was evidence for the existence and goodness of God.

Paley's detailed exposition of this argument in his book *Natural Theology* contained more than a dozen chapters of natural history, which provided the young Charles Darwin with a veritable textbook of examples of adaptation. When Darwin read *Natural Theology* as a student, while he himself was preparing to become a clergyman, he and his contemporaries accepted the argument from design as a rational basis for belief in a divine power. Essentially the same argument is still made today by creationist proponents of 'intelligent design'.

Design without a designer

Darwin boarded the Royal Navy's exploration vessel HMS *Beagle* in 1831 as a believer in Paley's theological explanation for adaptation. But during the five-year voyage Darwin observed variation within species between one island and another in the Galapágos Islands, and he began to doubt that species were fixed types, each separately created. He saw fossils of extinct animals in rocks near the coast of South America and thought about their similarity to living forms. The fossils and patterns of variation suggested to him that species changed, and that new species always descended from existing ones.

CANINE FRIENDS IN SICKNESS AND IN HEALTH

ABOVE Six breeds of dog, all descended through artificial selection from the grey wolf.
Top left to right: Ibizan hound, Boston terrier, cocker spaniel.
Bottom left to right: foxhound, greyhound and mastiff.

Darwin was so impressed by the variation among breeds of dog that he argued they could not have all been derived from the same wild ancestor. Thanks to modern genetic research, we now know that Darwin need not have been so cautious. The 400 or so breeds of dog are mostly the product of human selection from one wild species – the grey wolf.

Wolves were probably domesticated several times in different places, most recently at least 15,000 years ago. However, many of the modern breeds have been produced very much more recently, descended from just a handful of dogs whose ancestors had been under domestication for thousands of years.

The breeders of dogs and other domesticated animals carefully preserve their pedigree status by close inbreeding. Mating among close relatives, combined with artificial selection for particular characteristics such as coat colour, shape of the body and so on, reduces genetic variation among members of a breed and tends to increase the frequency of expression of genetic abnormalities. For example, a hereditary cancer of the kidneys occurs in German shepherd dogs; a kind of muscular dystrophy is common in golden retrievers, and Doberman pinschers are particularly afflicted by narcolepsy.

Because such pedigree dogs are genetically so alike, the genes that cause disorders in some individuals are easier to pick out when their genomes are compared with those of unaffected dogs of the same breed: the relatively small number of genes that are different between diseased and healthy dogs of a breed are the prime suspects. Since all mammals, dogs and humans included, share so much of their physiology, inherited from our common ancestors, the faulty genes that cause disease in dogs are frequently the same ones that cause the equivalent diseases in humans. Several hundred genetic disorders that affect both dogs and humans have been identified. Only because of our common evolutionary history is it possible to use dogs, or indeed other mammals such as guinea pigs or lab rats, as laboratory models for human disease in the search for treatments and cures.

However, it was not enough just to propose that evolution occurred. Others had proposed this before but failed to explain how evolution happens and, as a result, the idea had failed to take root. In *On the Origin of Species*, which Darwin wrote many years after the *Beagle* voyage, he said that the evidence he had collected might lead one:

> '*... to the conclusion that each species had not been independently created, but had descended, like varieties, from other species. Nevertheless, such a conclusion, even if well founded, would be unsatisfactory, until it could be shown how the innumerable species inhabiting this world have been modified, so as to acquire that perfection of structure and coadaptation which most justly excites our admiration.*'
>
> Darwin, C.R. (1859) *On the Origin of Species by Means of Natural Selection*

In other words, it is not enough to establish that evolution has taken place. We must also discover how it happened in a way that gives the appearance of design. Paley had made a case for design by God and if that argument was wrong, it must be answered with another explanation for how adaptation could be marvellous without also being miraculous.

BELOW Four pigeon breeds produced by selective breeding from the rock dove. Charles Darwin was a pigeon fancier and belonged to two pigeon clubs.

Darwin described evolution as "descent with modification" and his starting point in *On the Origin of Species* was to show how powerful selective breeding had been in creating many varieties of domestic animals. For example, so different are the various breeds of pigeon, which fanciers have bred from a single species, the rock dove, that an ornithologist would justifiably classify them as different species if told that they were wild birds. By analogy, Darwin then argues, natural selection can be just as powerful, and given enough time, that power can have dramatic effects.

> *Let it be borne in mind how infinitely complex and close-fitting are the mutual relations of all organic beings to each other and to their physical conditions of life.*

Today, we describe these mutual relations as ecology. Darwin continues:

> *Can it, then, be thought improbable, seeing that variations useful to man have undoubtedly occurred, that other variations useful in some way to each being in the great and complex battle of life, should sometimes occur in the course of thousands of generations?*

If we can selectively breed from variation found in wild species to produce new varieties, then nature can do the same thing.

> *If such do occur, can we doubt (remembering that many more individuals are born than can possibly survive) that individuals having any advantage, however slight, over others, would have the best chance of surviving and of procreating their kind?*

Any inherited variation that confers an advantage that causes its carriers to leave more offspring than the average will spread.

> *On the other hand, we may feel sure that any variation in the least degree injurious would be rigidly destroyed. This preservation of favourable variations and the rejection of injurious variations, I call Natural Selection.*
>
> Darwin, C.R. (1859) *On the Origin of Species by Means of Natural Selection*

Natural selection is a winnowing process in which favourable variants multiply at the expense of less favourable ones.

Even today, the only mechanism we know of that can produce long-term heritable adaptation is Darwin's mechanism. Natural selection has justifiably been called the best idea that anyone ever had. The reason that natural selection was such a revolutionary idea was because it explained how living beings that contain complicated structures, which look as though they must have been designed by an intelligent and skilled designer, can in fact arise without any supernatural intervention. This new concept turned on its head the argument that Darwin had read and accepted as a student. Now, the many wonderful biological mechanisms that Reverend William Paley described in *Natural Theology* were no longer evidence of divine, supernatural design, but of an entirely natural process operating without any guidance from above.

Natural selection

Darwin's argument for natural selection is, in essence, remarkably simple.

First, Darwin says, life is complicated. Organisms depend for their survival upon relationships with other species, such as the animals that eat them or the animals they eat and those that compete for the same resources. The close fit between different species provides examples of adaptation.

Then, he says, we know from the breeding of domesticated species that heritable variation is thrown up during the course of generations, and similar processes must happen in the wild too. Over thousands of generations in the wild, some of the variation that occurs is bound to improve an individual's success in what Darwin first called "the great and complex battle of life", and later in *On the Origin of Species*, "the struggle for existence".

This struggle arises because many more offspring are produced than can possibly survive: a large oak may produce millions of acorns in its lifetime, a herring produces millions of eggs, yet on average only one acorn or two fertilized eggs per generation can survive to adulthood or the land would be packed coast-to-coast with oak trees and the oceans would be stiff with herring. Darwin referred to the tendency of reproduction to produce a huge excess of offspring as superfecundity.

The struggle for existence winnows the better adapted from the rest, favouring any new variation that gives its bearer an edge in reproductive success. Variation favourable in the struggle for existence that is inherited by the offspring is perpetuated in descendents and thus success breeds success. Hence, natural selection increases the frequency of any hereditary changes that are favourable to success in the contest to be one of those very rare acorns, or the most fortunate pair of eggs that survive to form the next reproductive generation of oaks or herring.

The whole mechanism of natural selection is powered by just three things: variation, inheritance and superfecundity.

At the time Darwin published *On the Origin of Species* he was keenly aware of the many gaps in the evidence that were needed to establish his theory. He even devoted a chapter of the book to these difficulties, pointing them out to the reader. After a century and a half of research on evolution, not one of the difficulties remains.

Chief among the gaps that have now been closed in Darwin's theory was the problem of inheritance. It was obvious that children inherited some, but not all, of their parents' characteristics and that animals and plants with particular characters usually produced offspring that were similarly endowed – but how it worked, no one understood. Darwin devoted considerable energy and time to trying (unsuccessfully) to solve this problem. As it turns out, Darwin's greatest challenge was to become modern biology's greatest triumph: the science of genetics.

Second only to the problem of inheritance was the problem that Darwin was unable to cite any actual example of natural selection operating in the wild. Instead, as illustrated in the quotations above, he had to argue on logical grounds that it *must* occur, supporting his case by using the analogy of artificial selection in domesticated livestock and plants. Today, not only are numerous examples of natural selection known, but the genetic basis of the characteristics that are under selection is also understood in many cases. It took nearly a century from the publication of *On the Origin of Species* in 1859 for a clear-cut example of natural selection to be demonstrated, though the story really began with the growth of industrial pollution in Britain in the middle of the 19th century. The evolution of a condition now known as industrial melanism in the peppered moth is a story recounted in nearly every biology textbook, but recent research has impressively strengthened the evidence.

In unpolluted habitats, the 'typical' peppered moth (*Biston betularia*) is off-white in colour, generously flecked with black marks. In 1848, a pure black (melanic) peppered moth was captured in a garden in Manchester, England, which, at that time, was expanding rapidly and burning large quantities of coal that released black soot. Over the ensuing 16 years, the frequency of the melanic form greatly increased in Manchester and spread to other parts of the industrial Midlands of England; by 1900, it had been found throughout the country, with the exception of areas only slightly affected by air pollution in the western counties. These changes occurred so rapidly that they were noticed by many naturalists. In 1896, one of them suggested that the evolution of the melanic form was the product of natural selection, driven by the ability of the birds that prey upon moths to see them.

The peppered moth flies at night and rests on the boughs of trees by day where it is vulnerable to birds

ABOVE Typical (upper) and melanic (lower) forms of the peppered moth. Change in colour forms of this moth was one of the first examples of evolution by natural selection to be witnessed and tested by scientists.

that forage in daylight. The typical form of the peppered moth is camouflaged and very difficult to see when resting against lichens that encrust the bark of trees in unpolluted areas. However, most lichens are unable to grow in heavily polluted air and on dark, lichen-less branches, typical, off-white moths are highly visible. Melanic forms are cryptic against a dark background, but highly conspicuous against a white, lichen-covered background to predators like birds that hunt by sight. Bats also prey upon moths, but they hunt at night and locate their prey while on the wing using ultrasound, rather as radar is used to locate aircraft. Because bats do not search for moths by sight, they cannot select peppered moths on the basis of their colour. Thus, even if bats eat more peppered moths than birds do, it is only selective predation by birds and not random killing by bats that can produce adaptive camouflage.

The difference in colouration between the typical (that is, mottled) and melanic forms of the peppered moth is hereditary, controlled by a single gene. In the early 1950s, Dr Bernard Kettlewell (1907–1979) decided he would give up his medical practice to undertake research that tested the idea that birds act as agents of natural selection, causing the evolution of camouflage colour in the peppered moth. In a reserve for woodland birds in the industrial city of Birmingham, where 85% of the peppered moths were melanic, Kettlewell released large numbers of marked individuals of both types of moth and then recorded the proportions of each kind that were later recaptured at a moth trap. He performed the experiment in 1953 and again in 1955, and in both years, twice as many marked melanics as marked typicals were recaptured, indicating that in this polluted wood, typical mottled peppered moths survived less well than melanics. In other experiments, Kettlewell observed that birds did indeed prey upon more conspicuous types, as predicted.

He then repeated his mark–release–recapture experiment in an almost unpolluted wood in Dorset where 95% of moths were typical. Here he discovered that, as expected, the typical mottled moths were recaptured at twice the rate of the melanics that had been released. Here too, birds were eating more of the conspicuous type, but in this wood, it was the melanics that stood out against the branches upon which they settled, not the ancestral mottled moths.

Kettlewell's experiments were a powerful demonstration of Darwin's prediction that natural selection, acting through the agency of "mutual relations of all organic beings to each other and to their physical conditions of life", would produce adaptation. Since the 1950s, evolution of industrial melanism has been discovered in other species such as spittle bugs and ladybird beetles as well as in peppered moths in many parts of Europe and in the United States. Gradients in the frequency of melanics have been discovered, starting at high levels in industrial, polluted areas and diminishing with distance into the cleaner air of the countryside. Further experiments and observations with birds have confirmed the importance of predation as the agent of natural selection. Most convincingly of all, as anti-pollution legislation and the use of fuels other than coal and de-industrialization have reduced soot particles in the air significantly in Europe and the United States, the frequency of melanic peppered moths has also fallen.

It is perhaps ironic that Darwin's training for the clergy exposed him to the very evidence of adaptation in *Natural Theology* that he later re-interpreted to show that

LEFT The inscription on the bell-tower of the church at Downe in Kent where Charles Darwin lived for most of his life.

BELOW The sundial on the parish church at Downe in Kent.

divine intervention was not required to explain it. But, the argument from design did not originate with Reverend William Paley. Two thousand years earlier, the Roman Senator Cicero wrote:

> *When you see a sundial or a water-clock, you see that it tells the time by design and not by chance. How then can you imagine that the universe as a whole is devoid of purpose and intelligence, when it embraces everything, including these artifacts and their artificers?*

Though William Paley substituted a watch for Cicero's sundial, the more ancient timepiece did reappear. In the village of Downe in Kent where Darwin lived for most of his adult life, the village church has a sundial placed high up on the flint wall of the bell tower. The sundial is a resonant memorial to the man whose brilliant idea changed a view of the world that had lasted 2000 years, but whose time had come. The importance of that idea is difficult to over-estimate, for it implies that all organisms that have ever lived belong to one giant genealogy: the tree of life.

OPPOSITE The bell-tower of the parish church at Downe in Kent.

PEDIGREE OF MAN.

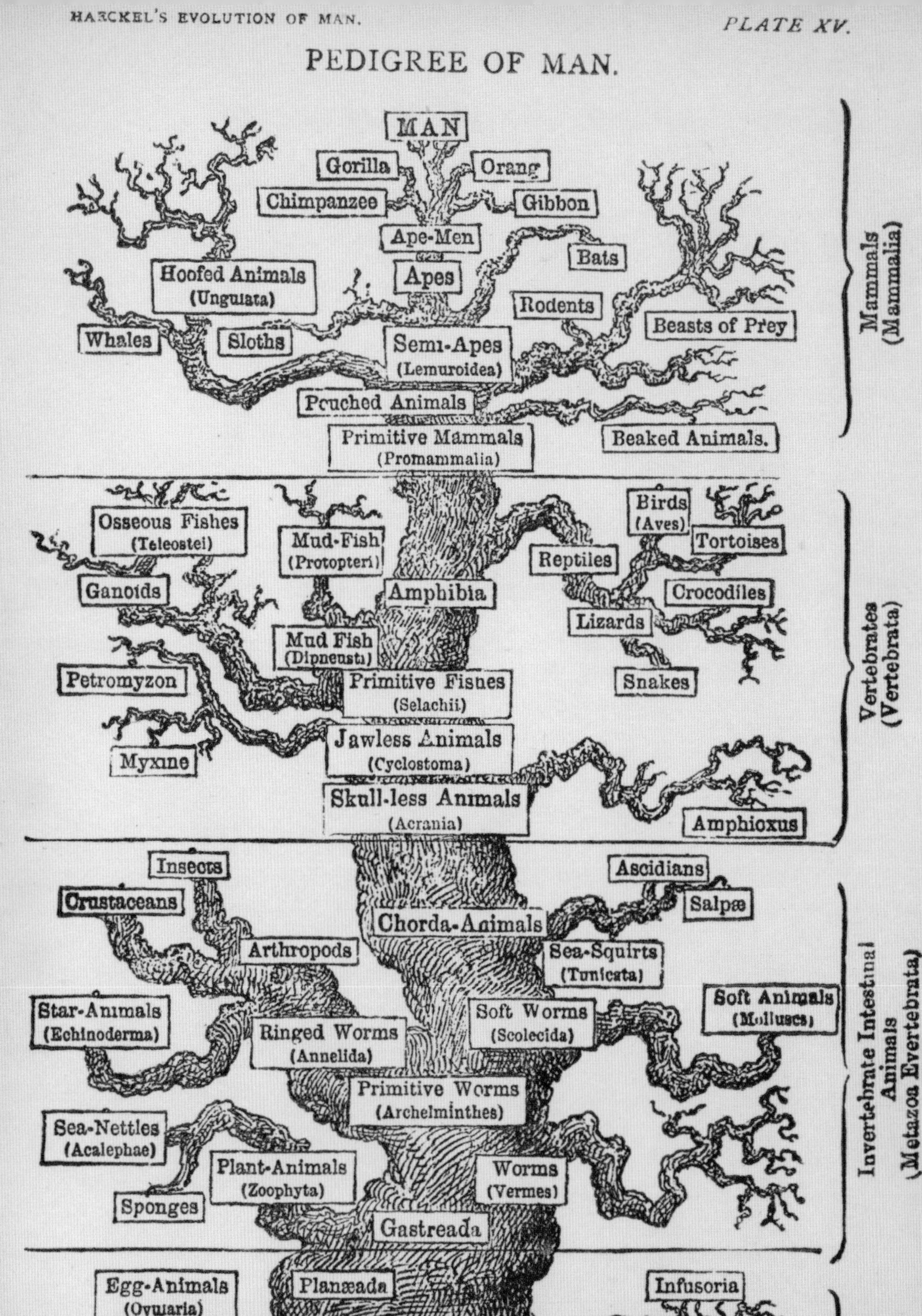

3 The tree of life

CHAPTER

RIDING THROUGH A BRAZILIAN FOREST IN 1832, the young Charles Darwin was awed by the exotic spectacle. He knew from his boyhood passion for collecting beetles about the seemingly boundless variety of living species, but the rainforest was a veritable naturalist's paradise.

LEFT The Brazilian rainforest – a naturalist's paradise, today recognized as one of the world's greatest 'hotspots' of biological diversity.

> ' APRIL 18TH. — *In returning we spent two days at Socêgo, and I employed them in collecting insects in the forest ... It is easy to specify the individual objects of admiration in these grand scenes; but it is not possible to give an adequate idea of the higher feelings of wonder, astonishment, and devotion, which fill and elevate the mind.* '
>
> Darwin, C.R. (1839)
> *Narrative of the Surveying Voyages of His Majesty's Ships Adventure and Beagle*

He was far from the first person to wonder at the rich diversity of life. A century before, the Swedish naturalist Carl Linnaeus (1707–1778) had set out to name all the species of life on Earth, in an attempt to tackle one of the great philosophical puzzles of the time – 'the plan of creation'. Despite naming some 15,000 species, Linnaeus only scratched the surface of the 30 million or so now estimated by some authorities to share the Earth with us. Nevertheless, biologists still use the Linnaean system of classification today because it succeeds in bringing a sense of natural order to the world's wealth of biodiversity.

OPPOSITE Darwin's recognition of evolutionary relationships between organisms gave a new, genealogical meaning to the idea of 'the tree of life', as expressed in this evolutionary tree figured by the German biologist Ernst Haeckel in 1874.

SCIENTIFIC CLASSIFICATION OF SPECIES

Each species is assigned to a grouping of similar species known as a genus, yielding a double-barrelled name (in Latin form and always italicized), such as *Quercus robur*, which is the common or pedunculate oak. Other species assigned to the oak genus *Quercus* include *Q. petraea*, the sessile oak, and *Q. ilex*, the holm or evergreen oak, as well as dozens of other species of oaks. (Note that generic names are usually abbreviated to a single letter after their first mention.)

Genera (plural of genus) are in turn grouped in families, families in orders, orders in classes, classes in phyla, and phyla in kingdoms. Such groupings are collectively known as 'taxa' (singular: taxon). Oaks belong to the family Fagaceae, which also includes the beech genus, *Fagus*. This family is placed in the order Fagales, within the class Magnoliopsida (also including magnolias, as its name suggests) and phylum Magnoliophyta, in the plant kingdom, Plantae. Taxonomic names above the level of genus are not italicized and start with a capital letter.

ABOVE Three species of the oak genus, *Quercus*: left to right, common oak (*Q. robur*), sessile oak (*Q. petraea*) and holm oak (*Q. ilex*).

The Linnaean system of classification was already well established when the young Darwin was collecting insects in the Brazilian forest in 1832. Yet he was to be the one who would convince the scientific world of a rational explanation for its hierarchical structure.

Cracking the mystery of species

Historically, three main hypotheses have been proposed to explain how the world's wealth of living organisms came into being.

- **SPECIAL CREATION**: species were individually created and remained fixed in character. Derived from the biblical Book of Genesis, this was the most widely held view until the publication of *On the Origin of Species*; it was also what the young Darwin initially believed as he voyaged around the world on the *Beagle* in 1831–1836.
- **TRANSFORMATIONAL EVOLUTION**: modifications acquired by individuals during their lives are passed on to their offspring, causing the progressive transformation of species. Especially associated with the French naturalist Jean-Baptiste Lamarck

(1744–1829), this idea was also favoured by most early evolutionists (including Erasmus Darwin). In fact the word 'evolution', from the Latin for 'unfolding', originally referred to individual development and only acquired its modern sense following the publication of *On the Origin of Species.*

- **VARIATIONAL EVOLUTION**: small heritable differences between individuals accumulate variously in separate populations over long periods, eventually giving rise to the larger differences between species and higher taxa. This idea combines descent with modification together with a branching pattern of diversification, analogous to a family genealogy (see overleaf).

Ideas on trial

Scientific hypotheses are like suspects in a 'whodunnit?'. Both science and detective work, properly conducted, depend upon weighing the evidence to eliminate suspects and to nail the perpetrator.

Two kinds of evidence allow us to test the hypotheses for how species arise: firstly, direct records of ancient life in the form of fossils; and, secondly, clues from living organisms that betray how they may be related. Testing involves predicting what we should expect to see if a given hypothesis is correct, in contrast to what other hypotheses would predict, followed by review of what the evidence actually shows. If, after repeated testing based on different lines of evidence, one hypothesis remains the only reasonable option, it can then be accepted as a sound scientific theory (effectively as 'fact').

The only testable prediction that the creationist hypothesis makes is that the fossil record should show all species to have remained fixed following their abrupt first appearance. If each one had been independently created, there would be no reason to expect to find evidence of relationships among different species. Special creation therefore makes no predictions about how species are related to each other, or why such relationships should exist. Our close genetic affinity with apes, for example, is certainly not predicted by creationism. Nor does the hypothesis even specify exactly *how* new species are created. This lack of explanatory power is a problem for creationism that we will come back to in Chapter 18 *The* science *of evolution.*

With regard to the timing of appearance of different species, early 19th-century geologists already knew enough about the fossil record to doubt the literal biblical story of the creation of all life in a single week. Accordingly, they developed more sophisticated alternatives involving creation over longer spans of time (see pp.32-33). Nevertheless, still having no idea as to the creative process for the multiple acts of creation that they proposed, they frankly admitted its mysterious nature:

> *'To what natural or secondary causes the orderly succession and progression of such organic phenomena may have been committed, we are as yet ignorant.'*

Owen, R. (1849), quoted in Rudwick, M. (1972) *The Meaning of Fossils*

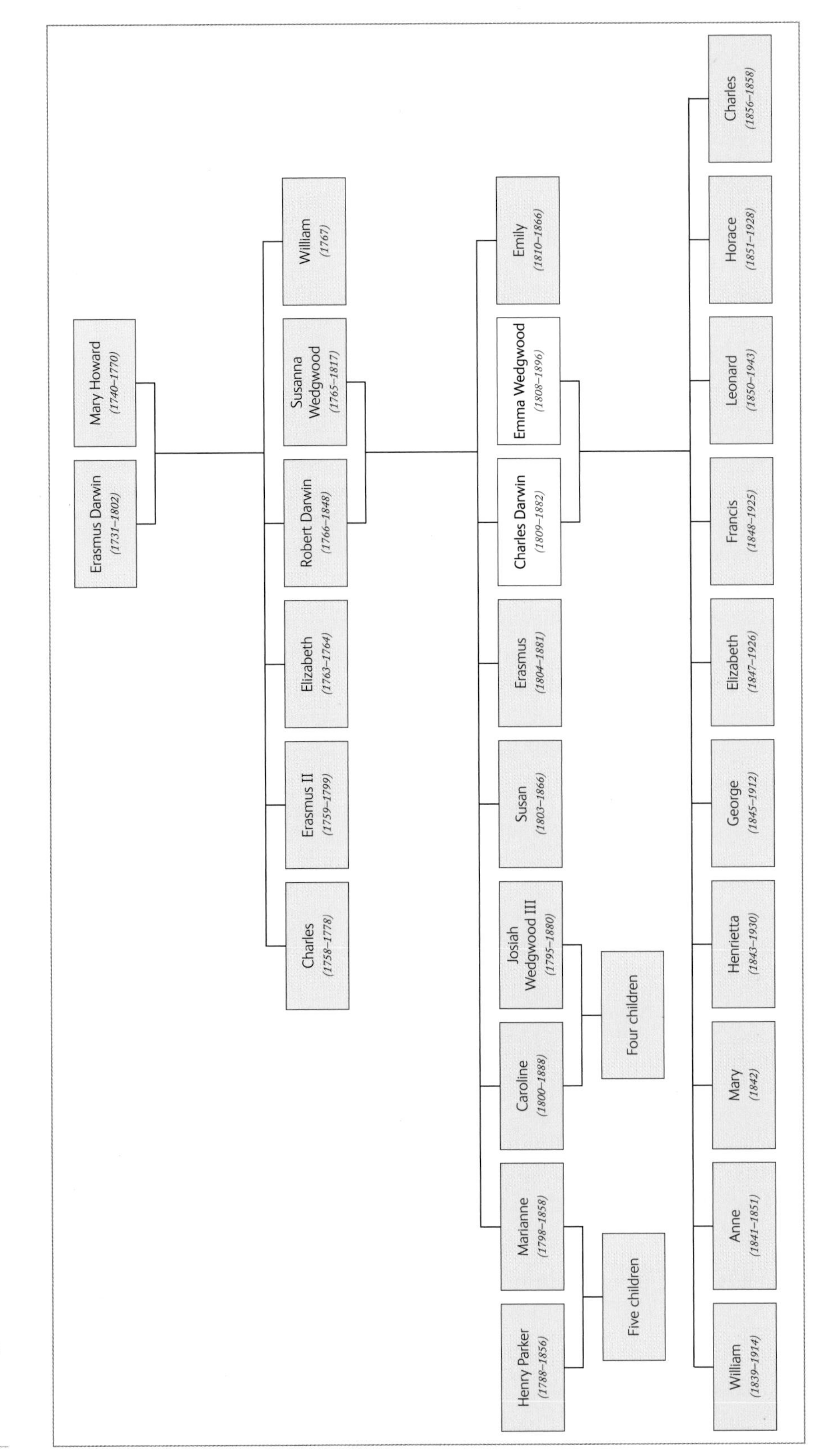

RIGHT Genealogy of the descendants of Erasmus Darwin and his first wife, Mary Howard down to the generation of Charles Darwin's children.

So what *do* the fossils show? In Darwin's day no examples of gradual evolutionary change in fossil species were known, which he blamed on gaps in the fossil record. Most sedimentary sequences do indeed consist of layers of rapidly deposited sediment separated by gaps representing long intervals of non-deposition or even of erosion of previous deposits. Think, for example, of the intermittent way in which rivers carry sand and mud onto their floodplains. Accordingly, the sampling of past populations as fossils entrapped in the sediment is, in most cases, extremely sporadic. But diligent collection, especially over the last few decades, has furnished many examples of continuous descent with modification, some marked enough to be recognized as transitions between species and even genera.

ABOVE Side views of modified tooth of *Arvicola* (above) and rooted tooth of *Mimomys* (below).

Most small rodents have huge populations and short lives which, combined with the excellent potential for preservation of their robust dentition, means that fossils of their teeth are widespread and abundant. This rich record has allowed scientists to describe some significant changes during the evolution, for example, of the water vole (*Arvicola terrestris*). The molar (cheek) teeth of this species have tall crowns with a zigzag arrangement of thick, hard ridges running up their sides, and much-modified roots. In most other mammals, the growth of rooted teeth is limited to the period before eruption from the gums. The specialized roots in water voles allow their high-crowned teeth to continue growing, replacing material lost as they are worn down in use. Fossil teeth from the water vole's evolutionary lineage can be traced back continuously to an ancestral form, called *Cromeromys savini*, in which more normal, though much-reduced roots are still present beneath the greatly extended crowns of the molar teeth. Moreover, an earlier lineage of successive species of another genus, called *Mimomys*, shows the

LEFT Water vole, *Arvicola terrestris*. The continuously growing molar (cheek) teeth of this rodent are adapted to its diet of coarse vegetation.

THE GEOLOGICAL TIMESCALE

ABOVE Unconformity at Siccar Point in Scotland observed by James Hutton. Truncated Silurian beds, below, are overlain by horizontal Devonian beds.

Younger sediments are deposited on top of older ones and this simple physical rule allows us to read the geological record of successive sedimentary layers (called strata) like the pages of a book. Over large regions, however, the picture can be much more complicated because discrete sequences of strata are often separated from one another by 'unconformities', where younger strata overlie the truncated ends of deformed older strata. This evidence persuaded the Scottish natural philosopher James Hutton (1726–1797) that there had been countless ages of sedimentary deposition each punctuated by an episode of deformation and erosion associated with mountain building. He concluded that "we find no vestige of a beginning, no prospect of an end".

Sedimentary successions and the fossils they contain testify to long periods of deposition in changing conditions. These facts forced geologists of the late 18th and 19th centuries to modify, and eventually reject, the biblical account of Earth history. At first, a series of discrete ages in the history of life (symbolic 'days' of creation) were proposed, each with its own separately created lifeforms, which then suffered catastrophic extinction before the next 'creation' appeared. In time, even this compromise was rejected when it was realized that different species had arisen and become extinct at different times. As more and more fossil evidence has been discovered over the last two centuries, geologists have built up the relative geological timescale of successive periods, grouped in eras. Calibration of this relative scale with absolute ages began in the 20th century, following the discovery of radioactivity, which allowed dating of rocks based on the ratios in them of unstable isotopes and their decay products. All available scientific evidence agrees that the Earth is exceedingly old, formed about 4600 million years ago, and that it has continuously supported life for the larger part of its history (Chapter 4 *First life*).

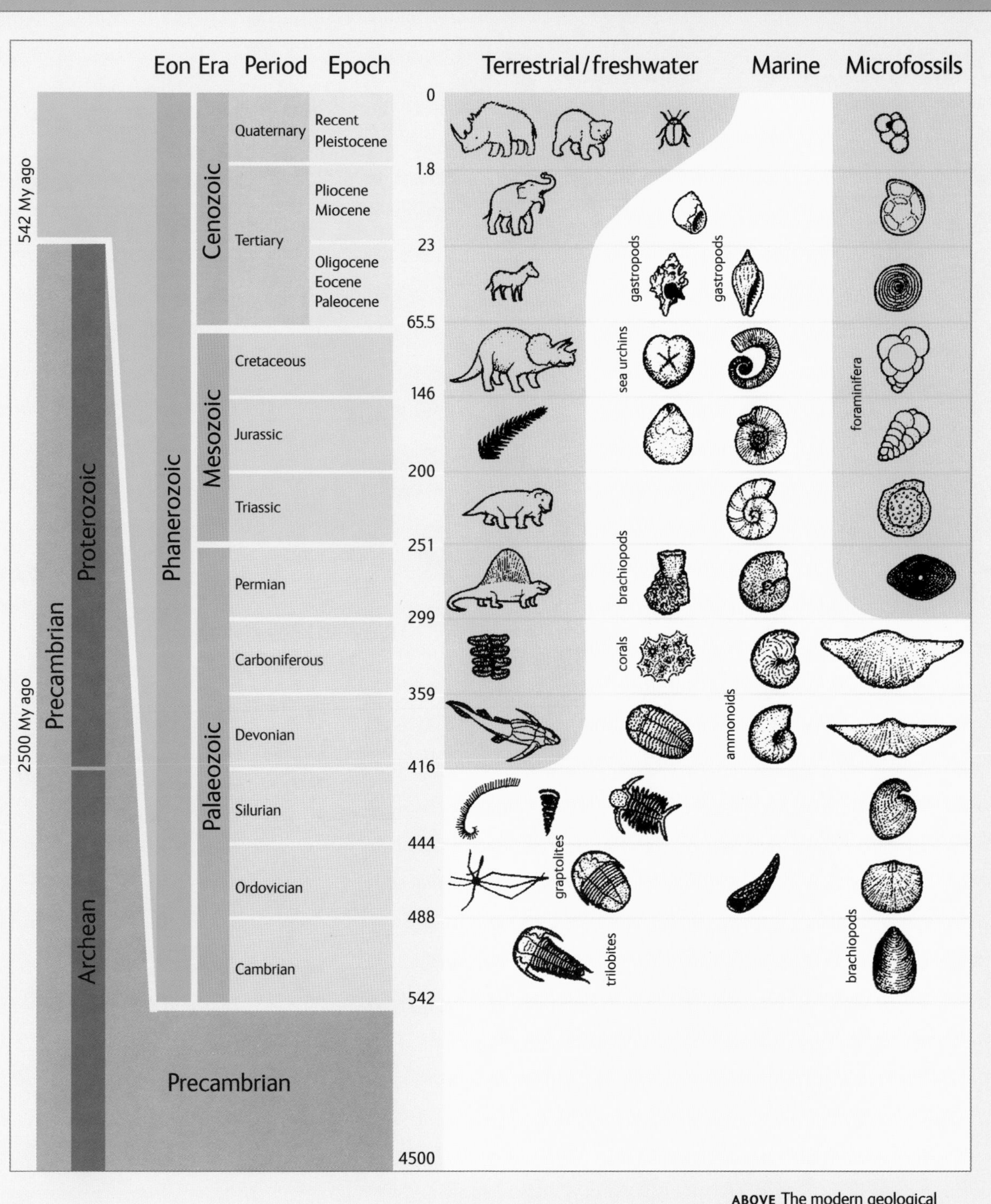

ABOVE The modern geological timescale. (Numbers indicate millions of years before the present.)

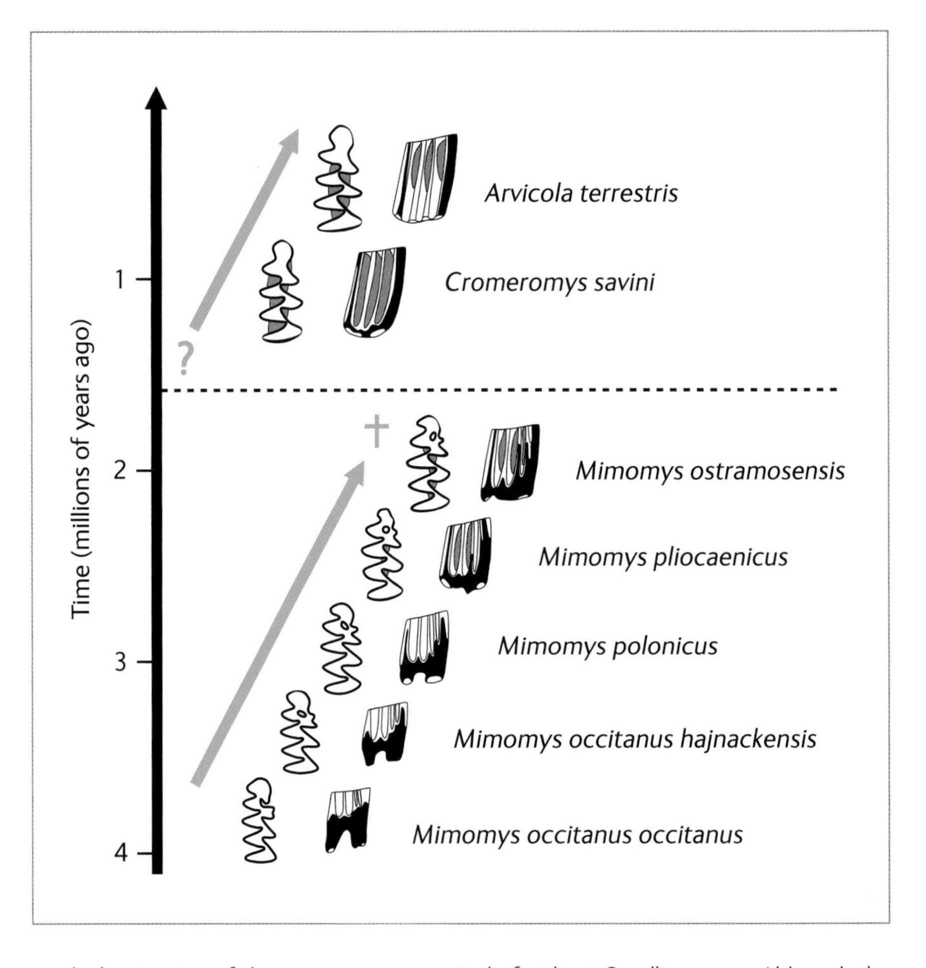

RIGHT Evolution of the molar teeth in water voles. Left, occlusal surfaces of teeth, showing zig-zag ridges of enamel, and right, side views, showing elongation of tooth crowns (white), in the *Mimomys* lineage, with subsequent loss of the root in *Arvicola*.

gradual extension of the crowns over a period of at least 2 million years. Although the last species of *Mimomys* was probably not directly ancestral to *Cromeromys savini*, it is likely that the latter was derived from a related form. So, the fossil evidence shows that species are not, after all, fixed in character, and can indeed be derived from others – even from forms different enough to be assigned to other genera. Hence the only testable proposition of creationism proves false, and we are left with evolution.

Darwin's evolving ideas

Soon after he returned from his voyage on HMS *Beagle*, Darwin began keeping a series of private notebooks on what he called the "transmutation of species". In 1837, Darwin emphasized three factors in his notes – adaptation, isolation and extinction – for all of which his experiences on the *Beagle* voyage had provided compelling evidence. Initially, he adopted a Lamarckian view on adaptation, which proposes that characteristics acquired during life can be transmitted to offspring. Influenced by the ideas of his grandfather, he supposed that the role of sexual reproduction was to furnish a range of offspring that varied in response to environmental conditions, as he scribbled telegraphically in his notes:

LEFT Darwin's iconic first sketch of an evolutionary tree, in his first private 'B' notebook on the transmutation of species.

> “*We see ~~living beings~~ the young of living beings, become permanently changed or subject to variety, according to circumstances, — seeds of plants sown in rich soil, many kinds, are produced, though new individuals produced by buds are constant, hence we see generation here seems a means to vary, or adaptation.*”
>
> Darwin, C.R. (1837) *'B' Notebook*

Usually, he suggested, interbreeding between such varieties kept species relatively uniform. However, remembering the diverse island species he had seen, he proposed that the accumulation of such modifications in isolated populations would allow divergence:

> “*According to this view animals on separate islands ought to become different if kept long enough apart with slightly differing circumstances. — Now Galapagos Tortoises, Mocking birds, Falkland Fox, Chiloe fox, — Inglish and Irish Hare –*”
>
> Darwin, C.R. (1837) *'B' Notebook*

Finally, he drew a sketch of a hypothetical evolutionary tree to illustrate how the minor differences that arise between species could accumulate with time and, aided by extinction of intermediates, yield the larger differences between higher taxa.

At this early stage in his evolutionary thinking, then, Darwin – like Lamarck – saw adaptive modification acquired during individual development as the driver of variation, with isolation and selective extinction determining the shape of the tree. Only later did he come to see *any* heritable variation as the raw material from which natural selection, acting on individuals, guides the evolution of adaptation (see Chapter 2 *Darwin's brilliant idea)*. Darwin himself never ceased to consider the effects of habitual use and disuse as one source of heritable variation. The science of genetics eventually put paid to that Lamarckian hypothesis, by revealing that such acquired modifications cannot be inherited.

The branching tree-like pattern of Darwinian evolution provided the key to the natural hierarchical grouping of organisms recognized in Linnaean classification, just as Newton's gravitational theory had at last made sense of Kepler's laws concerning the motions of the planets:

> “*Naturalists try to arrange the species, genera, and families in each class, on what is called the Natural System. But what is meant by this system? Some authors look at it merely as a scheme for arranging together those living objects which are most alike, and for separating those which are most unlike; … I believe that something more is included; and that propinquity of descent, — the only known cause of the similarity of organic beings, — is the bond, hidden as it is by various degrees of modification, which is partially revealed to us by our classifications.*”
>
> Darwin, C.R. (1859) *On the Origin of Species by Means of Natural Selection*

Branches and trees

The relationships among species are like the relationships among members of a family, but species have no birth certificates from which the evolutionary equivalent of a family tree, called a phylogeny, can be constructed. Instead, we must infer relationships and build phylogenies from the pattern of similarities among species. From a phylogenetic perspective there are three kinds of similarity. To understand the three kinds, we will simplify matters by assuming that new branches on the tree of life usually form by the splitting of one branch into two. (In this respect a phylogeny is *unlike* a family tree in which offspring always have two parents.) In a tree formed by splitting in this way, any trio of closely-related species at the tips of the phylogeny must consist of a pair of 'sister' species and a 'cousin' that is more distantly related to them. Now think of how we might be able to use inherited characteristics to distinguish sisters from cousins among the three species. A characteristic may be:

- shared exclusively by the sister pair because it has been derived from a novelty in their most recent common ancestor – an example of such a 'shared derived character' is the loss of the tail in the great apes and humans, in contrast to its well-developed state in nearly all monkeys, who in relative terms are evolutionary 'cousins';
- shared by the 'cousin' species and just one of the 'sister' pair, but be modified in the third, so qualifying as a 'shared primitive character' – in our trio of primates, an example of this kind of similarity would be an extensive covering of body hair, relative hairlessness being a derived condition in humans;
- evolved independently in different lineages, often performing similar functions – for example, both bats and birds have wings, though many other features of bats (such as live birth and suckling of the young) show them to be mammals, not birds, so the two groups must have evolved wings independently.

Only the first kind of similarity, that of shared derived characters, signals relative recency of common ancestry. Biologists today consider great apes to be allied more closely with humans than with monkeys on the evolutionary tree of life because of the many

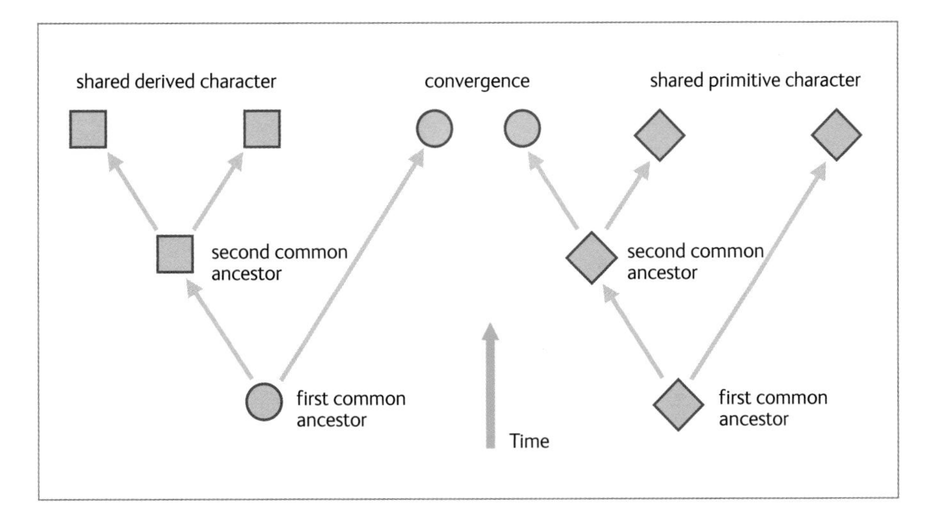

LEFT Three kinds of similarity between species. Symbols (circle, square and diamond) diagrammatically represent species' characters, those along the top as seen in living examples and those below, in their ancestors.

ABOVE Similarity due to convergent adaptations to nectar-sucking and hovering flight in hummingbird (left) and hummingbird hawkmoth (right).

shared derived characters (such as the reduction of the tail mentioned previously) that exclusively unite the first two of the trio. The relative hairiness shared by apes and monkeys, by contrast, merely reflects the recent reduction of body hair in humans and tells us nothing about which pair from the trio shares the most recent common ancestor. Likewise, evolutionary convergence due to adaptation to similar life habits does not reveal relationships.

If phylogenies are constructed on the basis of shared derived characters, we need a method to identify these. This is done by analytical procedures based on the principle that evolution is gradual and conservative. It follows that ancestral characters will be widely shared, while derived characters will tend to be confined to narrower groups, usually having a more recent common origin. Such tree-building methods work best when the number of characteristics that can be used is large. For this reason, mutations in DNA sequences are perfect material because they are so numerous and universal.

Enter the star witness – DNA

DNA is the molecule in which the genetic code is stored and transmitted. As the technology for reading the genetic code, called DNA sequencing, has improved to the point where it is now a routine process, comparisons of DNA sequences have provided a wholly new source of data for assessing evolutionary relationships that is independent of the classical reliance on the form and behaviour of organisms discussed above.

Although such analyses have prompted a fair amount of phylogenetic revision, they have strongly corroborated many well-established theories of relationship, such as that between the trio of humans, great apes and monkeys cited earlier. Moreover, as DNA is present in almost all living organisms, we can now confidently infer a single 'tree of life'. This tree turns out to have three major trunks (referred to as domains, or empires).

The Eucarya (or eukaryotes) is the domain to which we, and all the species around us that are big enough to see with the naked eye, belong. The shared derived character that defines the eukaryotes is having cells in which the DNA is packaged within a discrete membrane-bound nucleus. There are many single-celled eukaryotes, though more familiar are the larger multicellular forms that make up the plant, animal and fungal kingdoms.

The other two domains are the Eubacteria and the Archaea. These consist entirely of microbes whose minute single cells (most barely visible even under powerful optical microscopes) contain a single loop of DNA. The Eubacteria comprise the common bacteria, while the Archaea are typically limited to what we would regard as extreme environments that either lack oxygen or are highly salty, acidic or hot. These are the kinds of environments in which life may have first appeared on Earth.

Eukaryote cells contain numerous tiny membrane-bound bodies, called organelles, which perform certain key functions. Chief among these are mitochondria (singular: mitochondrion), which are present in almost all eukaryotes and provide energy by combining food molecules with oxygen, and chloroplasts, which conduct photosynthesis and are typical of plants. The presence of small amounts of DNA in mitochondria and chloroplasts that is closely similar to that in certain Eubacteria indicates that they were originally free-living bacteria, which became incorporated into ancestral eukaryote cells.

THE TREE OF LIFE (2008 EDITION)

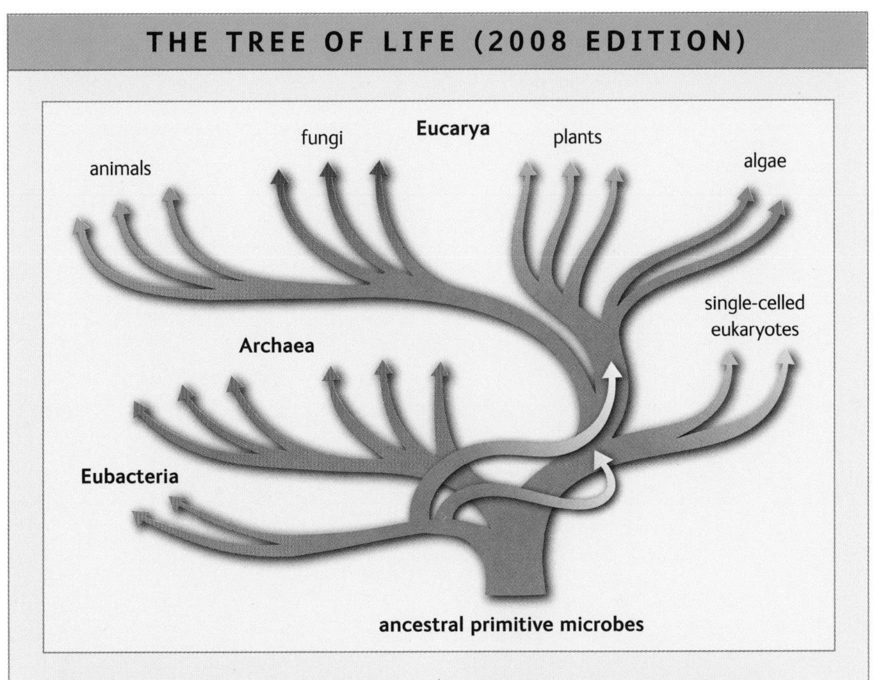

ABOVE The modern DNA-based 'tree of life'. All living species descend from a common single-celled ancestor that lived in the Precambrian and the branching pattern indicates evolutionary relationships inferred from comparisons of DNA sequences. Three major branches of the tree, Eubacteria, Archaea and Eucarya are indicated, as well as the main subdivisions of the Eucarya, including the three multicellular kingdoms that dominate macroscopic life. The two lines that run from the Eubacteria to the Eucarya represent the incorporation of precursors to chloroplasts (in plants) and mitochondria (in Eucarya), respectively.

CHAPTER 4 *First life*

> *... But if (and Oh! what a big if!) we could conceive in some warm little pond, with all sorts of ammonia and phosphoric salts, light, heat, electricity, &c., present, that a protein compound was chemically formed ready to undergo still more complex changes ...*
>
> Darwin, C.R. (1871) Extract from a letter to J.D. Hooker

BELOW A 'warm little pond' containing all the ingredients from which life might arise.

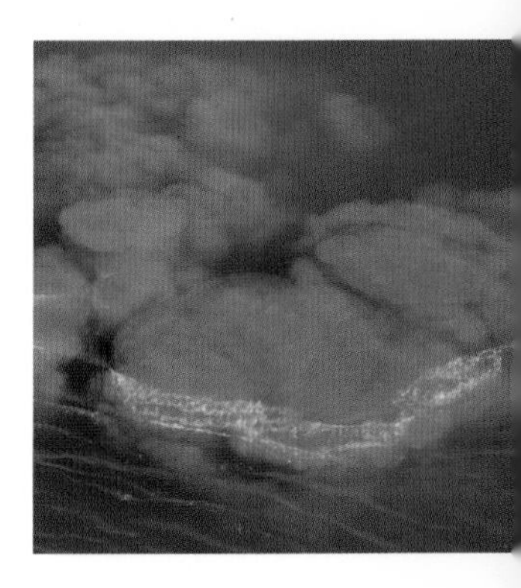

This description by Darwin of a 'warm little pond' is often taken to be a graphic and possibly prescient description of how he regarded the origin of life. However, if the next part of the letter is considered, we can see that Darwin was not just considering how and where life got going, but that he realized that the circumstances (that is, competition, timing) under which life originated were also important:

> *... at the present day such matter would be instantly devoured or absorbed, which would not have been the case before living creatures were formed ...*

The 'warm little pond' has been the framework for many of the investigations and proposals for how life originated that have been made since Darwin's day. So what have we learnt since then about how life arose? Where and when did it get going? These questions are addressed in this chapter. But before we consider the origin of life, we should first define 'life'.

The meaning of life

It may seem that 'what is life?' should be a fairly simple question to answer. Living things grow, reproduce and die. At the level of complex organisms (humans, elephants, oak trees), that is indeed the case, but at the inception of life there must have been very simple forms that crossed the line between the non-living and the living.

So what distinguishes something that is alive from something that is simply a collection of molecules? This question has occupied scientists, philosophers and theologians for many years, and there have been many attempts to define life and living. For present purposes, we define a living organism as something that is self-contained *and* is capable of self-replication with the transfer of genetic information. A virus does not qualify as living in this sense because it is not self-contained, and of course the first living thing could not have been a virus since it would have been unable to replicate without a host.

OPPOSITE Chemical reactions only occur if sufficient energy is available. It has been suggested that electricity generated by lightning bolts might have been the energy source for early life.

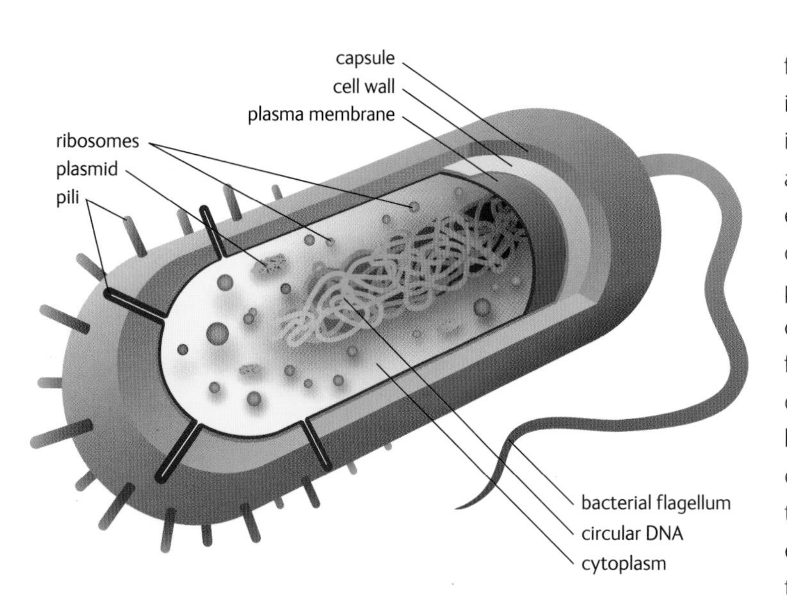

ABOVE The main features of a living bacterial cell. The outer layers isolate the cell from its surroundings and keep all the components of the cell together.

Each of these two criteria requires further investigation. 'Self-contained' indicates that the collection of molecules is somehow isolated from its surroundings and can exist independently of that environment. It implies a barrier, membrane or wall – not necessarily a rigid or non-porous barrier – that holds all the components together, preventing them from becoming separated. Keeping different molecules together is important because it enables them to react with each other at a rate fast enough for the entity to operate. Membranes regulate the exchange of chemicals between cells and their surroundings, and segregate compartments in larger, more complex cells. Membranes convert energy from external sources – food or sunlight, for example – into a biologically useful form. The most basic unit of a living system is therefore a cell, which is a collection of molecules held together by membranes – a bit like a small bag of fluid.

What does the second criterion, 'self-replication with transfer of genetic information', require? There are replicating systems that we would not regard as living: one example is the growth of a crystal from a solution. Imagine a rock pool as the tide goes out leaving a small quantity of seawater in the pool. As the water evaporates, salt crystals grow and spread across the surface of the rock. Each crystal has the same internal structure, if not

RIGHT A rock pool at low tide, with salt crystals produced by the evaporation of seawater. When the tide turns and the pool refills, the salt crystals dissolve, leaving no trace, or memory, of their presence.

LEFT A double-stranded coil of the DNA molecule showing how the coil can 'unravel' into two strands, each of which acts as a template for producing a new piece of DNA. This allows transfer of information from one generation to the next.

the same size, so it might seem that the first grain to crystallize replicated to produce a sheet of similar crystals. 'Information' is transferred, in the sense that subsequent grains grow along the faces of the first grains to form. The information flow is governed by the atomic structure of the sodium and the chlorine that make up salt. But can this information be transmitted? The tide turns, the pool fills up again with seawater and the salt dissolves. No 'memory' of the exact configuration of these particular crystals remains so that when the pool next drains out, the crystals grow according to the same principles but not to the identical pattern as on previous occasions, governed only by their atomic properties. Nothing has been 'learnt': there is no information transmitted from one generation of salt crystal to the next.

In a living system, information is transferred by the splitting and replication of DNA. The order of 'letters' on each strand of DNA is a genetic code that acts as an instruction, or template, for how the molecule re-forms after division. Although this code is almost always replicated faithfully, transferring information from one generation to the next, occasionally a mistake occurs and the code within one part of a DNA strand might become garbled. The altered sequence is a mutation and it is transmitted in future replications.

LEFT A coil of DNA following replication. The pattern of repeating units that hold the two strands together defines the genetic code that controls the function of the DNA.

In most cases, mutation is neutral in effect and has no consequences, because large parts of DNA molecules seem not to have any specific job. But sometimes the mutation is significant for the organism's structure or function, and can have either beneficial or deleterious consequences. These processes are part of evolution in action: a beneficial mutation imparts properties that allow better adaptation to an environment and the organism prospers to reproduce successfully. A deleterious mutation might impair organisms sufficiently for them to die out. If alteration in one small fraction of a DNA molecule occurs in a single instance of cell division, the consequences of that mutation may be perpetuated through many generations.

We see, then, that DNA transfers information that is essential to evolution. Coming back to our first criterion, 'containment', an entity must be self-contained for information to be transferred through generations. An additional constraint is that during replication, energy is consumed. So we should add 'using energy' to the definition of life. The most often quoted definition of life is that from Gerald Joyce:

> ‘*Life is a self-sustaining chemical system capable of undergoing Darwinian evolution.*’
>
> Joyce, G.F. (1994) *Origins of Life: the Central Concepts*

The phrase 'self-sustaining' encapsulates the requirement for energy and self-containment, while 'Darwinian evolution' implies transfer of information, inheritance and natural selection. The definition is imperfect: some entities are undoubtedly alive but do not conform to the definition (for example, worker bees, which are sterile so incapable of reproduction). However, this definition of life is useful as a guide when we consider how life arose on Earth.

Life's first spark

Life on Earth today, and as far as we can tell in the past, is based on DNA, a complex organic molecule that has the structure of a double helix. DNA is the product of a series of chemical reactions joining together simpler molecules (such as water, carbon dioxide and ammonia). The starting ingredients of life were a combination of material aggregated

RIGHT Halley's comet during its closest approach to Earth between 1985 and 1986. The nucleus of a comet is made of carbon-rich rock dust and ice, in which gases are trapped. Comets delivered volatile chemicals to the planet.

as the Earth formed, together with material added later by comets and asteroids. But the presence on Earth of the ingredients of life was not by itself sufficient for living organisms to form; there must also have been suitable environments in which the ingredients combined to produce more complex molecules.

A suitable environment includes an energy source capable of breaking and making chemical bonds, a surface on or in which chemicals can come together in sufficient concentration for reactions to occur on a reasonable timescale, and an extended period of stability of thousands, probably millions, of years, during which the reactants and products are sheltered from destructive forces while complex molecules form. This last condition was probably the main factor in delaying the formation of life on Earth, during the first few million years of its history, Earth's surface was unstable because of numerous impacts from comets and asteroids. The planet was also too hot to allow oceans to form, and so there were no suitable habitats in which life could become established.

BELOW The Earth was bombarded by comets and asteroids in the earliest phase of its history, keeping its surface molten. Eventually, the bombardment died down, allowing its surface to solidify.

Once Earth's surface had cooled and an atmosphere and oceans formed, it was possible for life to get going. But how? The mechanism by which simple molecules combined to form (eventually) DNA is not fully known, although there have been many suggestions. RNA, a similar but simpler molecule, must have evolved before DNA. In the early 1950s, Stanley Miller, a PhD student at the University of Chicago, carried out a series of what were to become classic experiments that influenced theories on the origins of life for decades.

Miller passed electric currents through mixtures of gases (ammonia, methane and hydrogen) and found that the tar-like deposit so formed consisted of various more

BELOW Experiments undertaken by Miller in the late 1950s, used pulses of electricity to fuse simple gas molecules together to produce a solid deposit of more complex organic molecules.

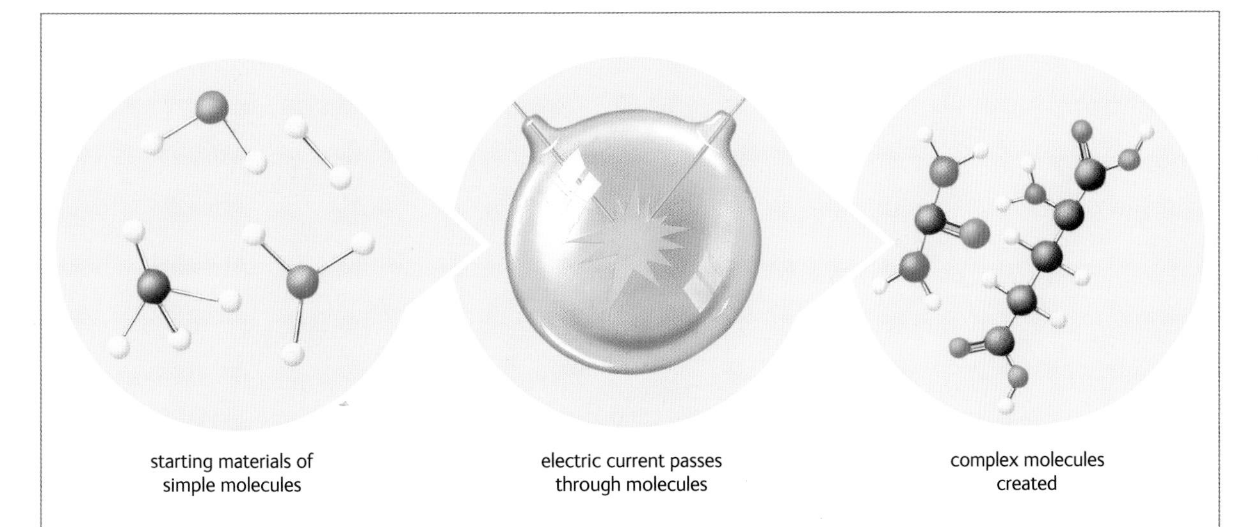

complex organic molecules that had been built up from the simple gas molecules. From these experiments, it was assumed that the first steps in chemical evolution must have taken place in the atmosphere, which (in the 1950s) was thought to contain substantial amounts of ammonia, hydrogen and methane. In such an atmosphere, radiation from the Sun, especially ultraviolet, and electrical discharges from lightning could provide the energy necessary to produce a wide array of molecules that together formed living cells.

Unfortunately for this theory, recent research shows that the early atmosphere was probably mostly nitrogen and carbon dioxide, and solar radiation is just as likely to destroy organic molecules (sunlight fades organic dyes and can kill living cells, as in sunburn) as it is to provide an energy source for their construction. So the hypothesis that the first stages of chemical evolution took place in the atmosphere was discarded.

It is now thought that rocks and minerals played an important role in the production of complex molecules. For simple building blocks to react further to produce larger molecules, especially polymers (chains of similar molecules linked together), the reactant molecules must be quite concentrated. The solid but rough and porous surfaces of many rocks or clay particles may have provided such concentrating environments, which were also capable of protecting the products of the reactions. Suppose that reactions on solid surfaces produced a rich mixture of biological polymers of the kinds found in living cells today. The mixture would still be far from being living cells, but it is now accepted that porous rocks could have been an essential step in the production of complex molecules and in the early stages of the evolution of life.

Earliest signs

Once we have identified a mechanism for how life might have started, it then becomes possible to deduce from the record that is preserved in rocks how early in Earth's history life began. This record can either be chemical (the preservation of traces of chemicals involved in the formation of life) or as fossilized organisms. Unfortunately, however, the record is incomplete and difficult to read, leading to uncertainty as to precisely when life first arose on Earth.

The unambiguous fossil record of living organisms stretches back 2.1 billion years ago or so, to the first stromatolites. Stromatolites are fossilized microbial mats formed by filamentous bacteria and algae alternating with layers of lime sediment trapped by them, like dust in a carpet. Although structures that look like stromatolites have been reported from 3.5-billion-year-old rocks in Australia, these examples are controversial. Nevertheless, some filamentous remains of possible bacterial origin have also been described from 3.2-billion-year-old deposits formed in hydrothermal vents on the sea floor.

Organisms have left chemical traces in rocks that are older than 3.2 billion years. As organisms degrade, the chemical residues of their cells can become trapped in sediments. Analysis of rocks formed from the sediments can identify the types of organisms that were present. So, molecules characteristic of dead, decayed eukaryotes, as well as

LEFT Section across a fossilized stromatolite from the 2,000-2,500 million-year old Precambrian rocks of Eastern Siberia. The section is about 230 mm (9 inches) in length.

bacteria, are found in rocks that are 2.7 billion years old, while 3.4-billion-year-old rocks contain the characteristic decay products of bacterial organisms. We infer, then, that life had arisen on Earth by some 3–4 billion years ago – that is, perhaps less than a billion years after Earth formed. That is not to say that life did not exist earlier, just that we have, as yet, found no record of it.

Earth's atmosphere today is 21% oxygen, but when life first evolved the atmosphere was essentially devoid of the gas that is now essential to the survival of most living organisms. The switch from an oxygen-less to an oxygen-containing atmosphere occurred somewhere between 2.5 and 2 billion years ago and was itself the result of evolution.

BELOW Thin section through a 2 billion-year old rock from Canada, showing the fossilized remains of blue-green algae, some of the earliest life found.

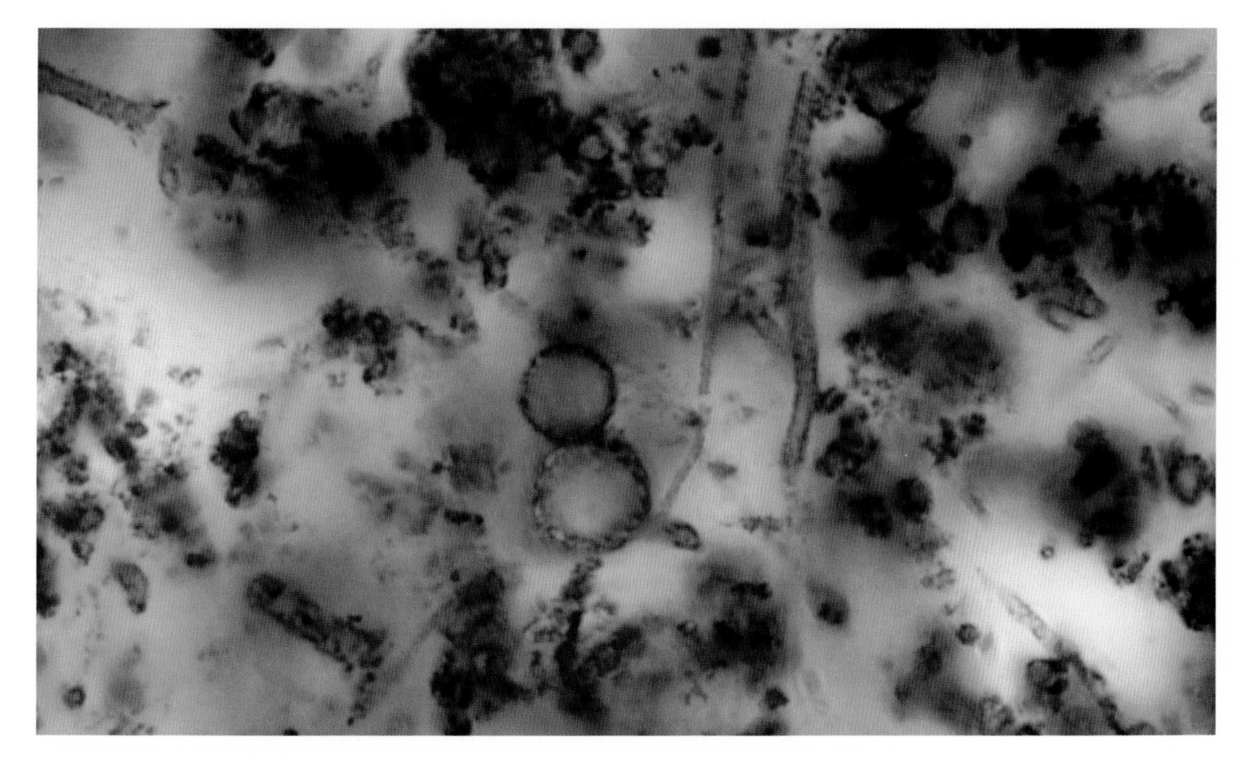

Oxygen was a waste product of photosynthesis by cyanobacteria and these organisms were responsible for this gas accumulating in the atmosphere. All plants today produce oxygen as a result of this same process, which occurs in their chloroplasts. These organelles are descended from an ancient evolutionary event when a eukaryote cell captured a cyanobacterium.

RIGHT Earth's atmosphere has not always contained oxygen – levels have gradually built up over the planet's history. As the amount of oxygen increased in the atmosphere, the complexity of surviving organisms also increased.

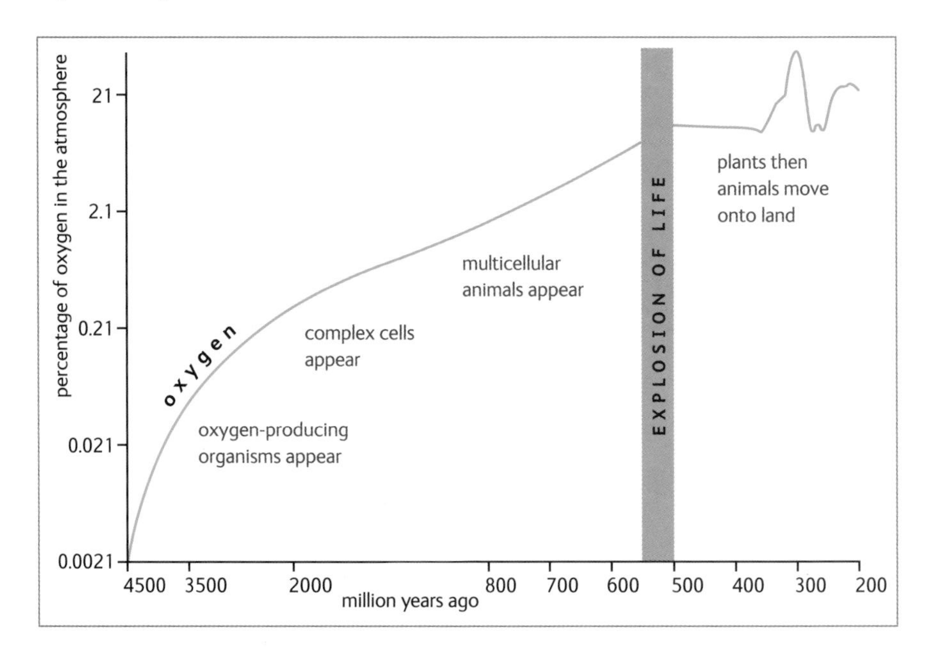

Life in hot water

Stromatolites like those found in ancient Precambian rocks are still forming today, in warm, shallow sea water. By analogy, then, it is possible that life first arose in warm waters, shallow enough to enable sunlight to penetrate and photosynthesis to occur. However, we now know that living systems based on chemical energy (as distinct from photosynthetic energy, the basis for life powered by sunlight in green plants and algae), exist in various other places on Earth, some of which are never exposed to sunlight.

Molten rock from deep in the Earth emerges to form new crust at spreading centres along ocean floor ridges. Hydrothermal vents were discovered close to these areas in the late 1970s. The vents, or 'black smokers', are hot springs where super-heated water (350–400°C, 660–750°F), rich in dissolved gases including hydrogen, methane and hydrogen sulphide, shoots up from the sea floor. Where the hot water meets the cold, relatively oxygen-rich sea water, an instant chemical reaction with hydrogen sulphide forms insoluble sulphides that colour the water black. The precipitated sulphides build up rapidly to 'chimneys' reaching heights of several tens of metres (tens of yards).

Scientists were surprised to find that, despite the depth and darkness, these parts of the ocean floor are home to a unique collection of animals including fish, mussels, crabs and tubeworms, some remarkably large. These animals can withstand the heat and chemicals that are toxic to most organisms, and feed on the microbes that flourish there.

ABOVE Giant tubeworms live around hydrothermal vents on the ocean floor, where they can grow up to several metres in length. They survive by feeding on bacteria that draw their energy from chemical reactions rather than photosynthesis.

LEFT Modern-day stromatolites at Shark Bay, Western Australia.

The discovery of such communities of microbes and animals based on chemical energy rather than photosynthesis has raised the possibility that life may not have arisen in sunlit surface waters. Discovering living communities based on the utilization of chemical energy has also given impetus to the search for life in other deep oceans, especially on Jupiter's satellite, Europa, where liquid water is thought to form oceans below a mantle of ice.

ABOVE Artist's impression of a pair of extrasolar planets orbiting a sun-like star.

As we have come to understand more about the range of life on Earth, we have seen that organisms inhabit a wide range of environments. The least complex forms of life (Eubacteria and Archaea) exist in a wide diversity of habitats. For example, some species of microbial life colonize places with extremes of temperature (–20°C to +110°C, –4°F to +230°F), acidity and alkalinity, salinity and radiation. Knowing that a 'warm little pond' is no longer a prerequisite for the origin and evolution of life allows us to explore beyond the Earth for signs of life.

Extraterrestrials

This book is concerned with the evolution of life on Earth, but it is well worth considering whether or not life might have arisen in other locations beyond Earth. If we base our expectations of extraterrestrial life on what we have inferred about the origin of life on Earth, there are several places within the solar system that could provide suitable habitats in which life might have evolved, or might do so in future. The most likely locations within

RIGHT Part of Mars' surface, showing features that are thought to have been produced by water at some time in Mars' past history. Water has now disappeared from the surface, but it is presently locked up in ice at the polar caps, and also in ice below the surface.

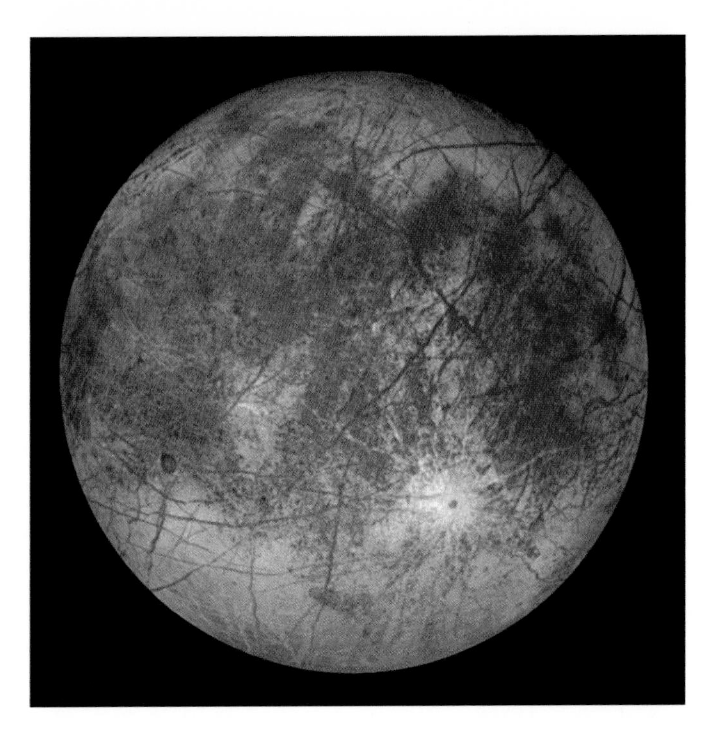

LEFT Jupiter's moon, Europa. Its icy surface is thought to cover a deep ocean, on the floor of which might be hydrothermal vents like those found on the Earth's ocean floor.

the solar system are the planet Mars and the satellite of Jupiter, Europa. The former has many indicators that water was there in the past, while Europa is inferred to possess a deep ocean of water below its thick icy surface. Both bodies are the focus of robotic space missions by international teams. There is thus the potential for us to study the evolution of life beyond our own planet.

It is a continued homage to Charles Darwin's foresight and genius that more than 130 years after he wrote to Joseph Hooker of energy-induced chemical reactions in an aqueous environment, his scenario is still one of the main hypotheses for how life arose from simple chemical building blocks. The discovery of micro-organisms on Earth that colonize habitats in environments of extreme heat, cold, acidity and radiation encourages the view that there may be life elsewhere. There is an abundance of organic materials and water within the solar system, and the potential for their occurrence on planets around other stars. But even if microbial life is widespread within the solar system or galaxy, there is no evidence, yet, that it has evolved into complex, intelligent beings of the kind that evolved in Africa.

5 CHAPTER *African genesis*

WHEN SOAPY SAM MET DARWIN'S BULLDOG across the debating chamber at a meeting of the British Association in Oxford in the summer of 1860, the encounter represented the first public contest of ideas between the new theory of evolution and established religion.

Bishop Samuel Wilberforce (Soapy Sam) asked Thomas Henry Huxley (Darwin's Bulldog), in essence, whether he claimed descent from an ape on his father's or his mother's side. Huxley was subsequently credited (wrongly) with replying that he would rather be an ape than a bishop. The precise words used were not recorded at the time, and several versions exist. However, the central point of their debate – the descent of humans from apes – is clear. Yet all that Darwin actually said in *On the Origin of Species* about human origins was in the third to last paragraph of the final chapter:

> *'Light will be thrown on the origin of man and his history.'*
>
> Darwin, C.R. (1859) *On the Origin of Species*

ABOVE Samuel Wilberforce (1805 – 1873) was Bishop of Oxford and later Winchester, a member of the House of Lords and, like Darwin, a Fellow of the Royal Society. He somehow acquired the nick-name 'Soapy Sam' perhaps because of his manner.

Today, we don't just regard humans as descended from apes; we share a common ancestry with the other apes, for we are apes ourselves.

Darwin did not set out his ideas on the place of humans in evolutionary history until 1871 in his book *The Descent of Man*. While he was working on his books, the first hard evidence of the human lineage was being discovered.

In the beginning ...

The African continent, ancestral homeland to the human species, is separated from the Iberian Peninsula of Europe by a narrow sea way that connects the Atlantic with the Mediterranean. Guarding the strategically important narrows between the two continents is the tiny British colony of Gibraltar, and it was on that rock that fossil evidence of human origins was discovered.

The Gibraltar Scientific Society met on 3 March 1848 in the Garrison Library on the Rock. The members were mostly, if not exclusively, serving soldiers and the secretary was Lt Edmund Flint, RA. In the minutes of that meeting, appears the following:

> *'Presented a Human Skull from Forbes Quarry, North Front, by the secretary.'*

OPPOSITE Thomas Henry Huxley (1825-1895), often referred to as 'Darwin's Bulldog', was a personal friend of Charles Darwin and a key exponent of his scientific work. As a fine lecturer and eminent scientist, he was the ideal person to champion the views of the rather more reticent Darwin.

ABOVE The Gibraltar skull from Forbes quarry.

With hindsight, this was a momentous occasion. Lt Flint died of apoplexy in 1857, and the details of the circumstances of the find, including the precise location, died with him. Nothing more was heard of the skull until another army officer, the governor of the military prison, included it in a consignment of fossils that he sent to George Busk at the Royal College of Surgeons in London in the summer of 1864. Busk, working with Hugh Falconer (Vice President of the Royal Society), made the connection between the Gibraltar skull and the finding of a skull cap and limb bones in the Neander valley in Germany, announced in 1857. The skull remains one of the finest and most complete skulls of a Neanderthal ever found, though if it had been recognized at the time of its discovery we should now be referring to the Gibraltar species rather than the Neanderthal species. Writing to J.D. Hooker in the evening on 1 September 1864, Charles Darwin wrote:

> *'Both Lyell and Falconer called on me and I was very glad to see them. F. brought me the wonderful Gibralter skull.'*

So, by the time that Darwin completed *The Descent of Man*, he had certainly seen one significant human fossil. In the book, he wrote:

> *In each great region of the world the living mammals are closely related to the extinct species of the same region. It is therefore probable that Africa was formerly inhabited by extinct apes closely allied to the gorilla and chimpanzee; and as these two species are now man's nearest allies, it is somewhat more probable that our early progenitors lived on the African continent than elsewhere.*
>
> Darwin, C.R. (1871) *The Descent of Man*

Considering the evidence available to Darwin was so meagre, his cautious statement about the probable African origin of the human species seems extraordinarily prescient. Since then, evidence from fossils has confirmed the African origin of the human lineage and applying our modern knowledge of inheritance, genes and molecules has reinforced the story told by the bones.

The 'prime' animals

Our ancestral line emerges from within the primate group – broadly, the lemurs, lorises, galagos, tarsiers, monkeys and apes. There has long been an assumption that our human species represents the pinnacle of an evolutionary line and thus humans and their close relatives formed a 'prime' group. Although the idea was discarded many years ago by science, it is surprising how long-lived it has been within our culture.

What features define a primate? It isn't easy to select a clear-cut set of characters and there are often exceptions. The hands and feet are capable of grasping, with opposable thumbs and sometimes also opposable toes. Nails replace claws in most primates. The centre of gravity of the body is close to the hindlegs, which almost always dominate in locomotion, although the arms are strong, as is the tail in New World

BELOW Gibbon, chimpanzee and tarsier are all primates and like these animals our primate ancestors were also tree-dwellers.

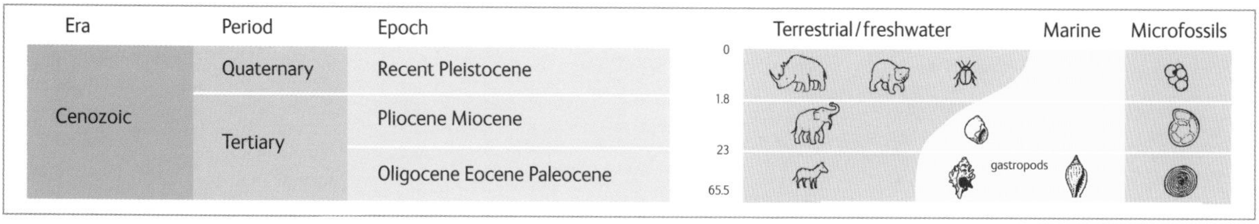

Era	Period	Epoch
Cenozoic	Quaternary	Recent Pleistocene
	Tertiary	Pliocene Miocene
		Oligocene Eocene Paleocene

Terrestrial/freshwater
Marine
Microfossils
0
1.8
23
65.5
gastropods

ABOVE The geological timescale of the Cenozoic era, during which primates evolved.

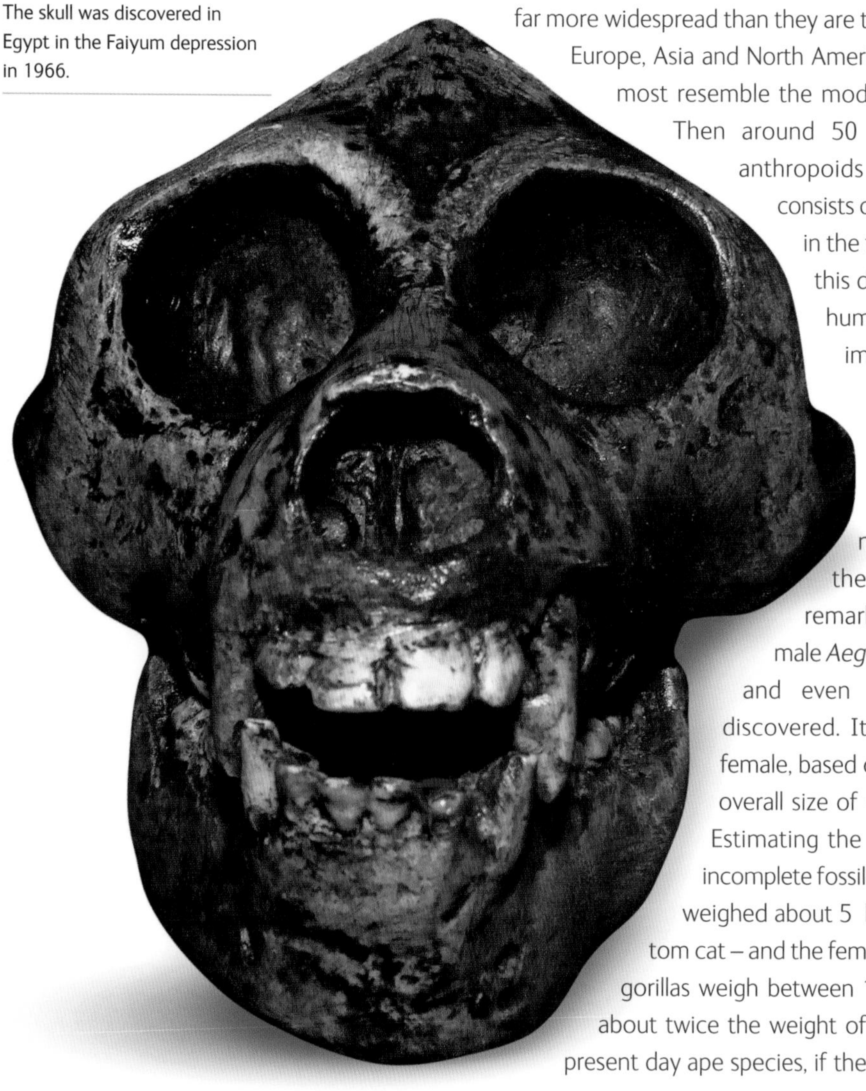

BELOW A cast of the skull of a male *Aegyptopithecus zeuxis*. The skull was discovered in Egypt in the Faiyum depression in 1966.

monkeys. Styles of locomotion vary, and include vertical clinging and leaping (tarsiers), knuckle walking (chimps), swinging (gibbons) and walking (humans and chimps). The snout is shortened and the eyes face forwards, allowing stereoscopic vision. Proportionately, the brain is larger, with an increased area devoted to visual processing and a reduced area devoted to smell. If you consider how well you match these definitions, you'll find quite a good fit, with the exception of grasping toes, which humans have almost lost as a flat foot with short toes became advantageous for walking. So with these primate features in mind, let us journey back in time to the African genesis of the human line.

In the Eocene epoch, around 55 million years ago, the primates were far more widespread than they are today, and there are fossils from Europe, Asia and North America as well as Africa. The fossils most resemble the modern lemurs and their relatives. Then around 50 million years ago, the first anthropoids (a subgroup of primates that consists of the monkeys and apes) appear in the fossil record in North Africa and this date really marks the root of the human line. The name anthropoid implies that they are man-like.

It appears that the first anthropoids lived in tropical forests, although the area where their fossils are found is now arid. A famous example of these early anthropoids was a remarkably well-preserved skull of a male *Aegyptopithecus*. In 2007, a smaller and even better preserved skull was discovered. It is almost certainly that of a female, based on the analysis of the teeth and overall size of other less well-preserved skulls. Estimating the original body weight from an incomplete fossil isn't easy, but the male probably weighed about 5 kg (11 lb) – about the size of a tom cat – and the female half that. Modern adult male gorillas weigh between 140 and 200 kg (300–440 lb), about twice the weight of mature females. Judging from present day ape species, if there are size differences between

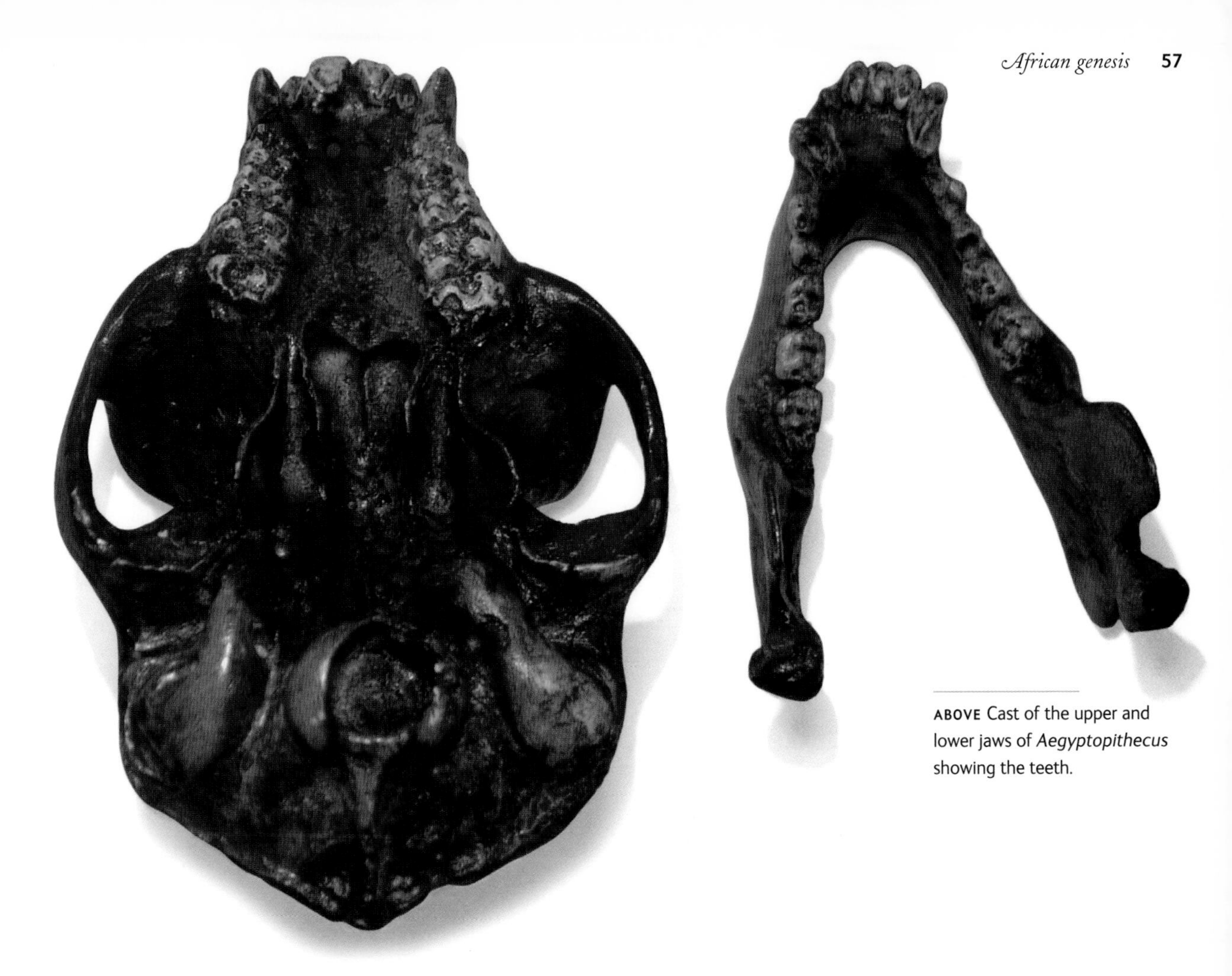

ABOVE Cast of the upper and lower jaws of *Aegyptopithecus* showing the teeth.

males and females there is likely to be a complex, intimate social structure, so maybe *Aegyptopithecus* lived in troops. The brain case of *Aegyptopithecus* has been studied with a CT scanner and the images show that the visual cortex is very well developed. The skull has large eye sockets that face forward, allowing binocular vision.

Jaws and teeth are the commonest fossils of primates as they are the parts that are most resistant to damage. Teeth can give us information about the diet of an animal and, perhaps more importantly, its evolutionary relationships. Adult humans have two incisors, one canine, two premolars and three molars on each side of both the upper and lower jaw. So do the apes, and the monkeys from Africa, Asia and Europe – the so-called Old World monkeys. So too does *Aegyptopithecus* and this shared array of teeth indicates a close relationship. However, *Aegyptopithecus* pre-dates the divergence between apes and Old World monkeys and is close in time and relationship to the first anthropoid.

For a period in the Miocene, apes flourished. They were large-brained, large-bodied animals and they spread through the forests of Africa, Europe and Asia. But about 10 million years ago, they began to die out as competition from monkeys increased. Today, there are only six species of great ape left, including humans (gorilla, two orang-utans, two chimps and humans), though the expansion of the human ape line has pushed the other five species to the brink of extinction. The reason for the monkeys apparently out-

ABOVE AND LEFT Four of the six surviving great apes, from far left gorilla, orang-utan, bonobo and chimpanzee. Humans are not shown and there are two species of orang-utan.

competing apes has to be largely a matter of conjecture, but a possible selective pressure was climate change, as the decline in the apes coincided with a period when the climate got cooler and drier. Monkeys with a greater flexibility in diet coupled to a tolerance of a wider range of climate conditions began to dominate and most apes eventually survived only where there were hot, wet forests in Africa and Southern Asia. In Africa the increase in aridity gave apes able to forage on the ground a selective advantage, and at some point less than 10 million years ago ground-living apes with a distinctively different anatomy appeared. From them developed the human line; an African genesis for what may well turn out to be the last ape lineage on Earth.

So, the idea of a ladder of progress culminating in modern humans doesn't stand up to examination. What we see in the Miocene is, as Stephen Jay Gould describes it:

> '*... the record of declining diversity in an unsuccessful lineage that then happens upon a quirky invention called consciousness.*'

The point at which the human apes diverged from the remaining apes is, for us humans, a critical point in our ancestry. We know where, in general terms it occurred – Africa – but when did it occur and what did the earliest hominid ape look like? These are questions that are begging for answers. There are two lines of evidence that we can examine: fossil and molecular.

History from molecules

The DNA molecules that carry our genetic code are present in every cell of the body. Within the cell, there are two places that you can find DNA. Nuclear DNA carries the bulk of the genes and is found, as its name suggests, in the nucleus of the cell. The DNA of a few genes is found outside the nucleus in small structures called mitochondria (Chapter 3 *The tree of life*) that house energy conversion processes.

Mitochondria are often described as the powerhouses of the cell and in an active tissue, like muscle, they are abundant, providing lots of copies of their DNA. Their significance for unravelling history is that their DNA carries the code for a small number of genes with a restricted role, which – in most animals and plants – are inherited only through the female line, and this makes ancestry easier to trace. Both eggs and sperm contain mitochondria, but when a sperm fertilizes an egg, the mitochondria that it carries are usually destroyed by the egg, or in some cases left outside with the tail. In sexually reproducing organisms, the mitochondrial DNA is normally inherited exclusively from the mother. So while each parent contributes 50% of the nuclear DNA to each offspring, there is no mixing of mitochondrial DNA (mtDNA) and it passes unchanged to the offspring. Change does take place over time, however, as mutations occur randomly in the letters of the genetic code.

BELOW The ancestral tree of the Old World monkeys and apes, derived from mtDNA studies. The figures in the circles represent the best estimate of the date at which each pair of branches diverged.

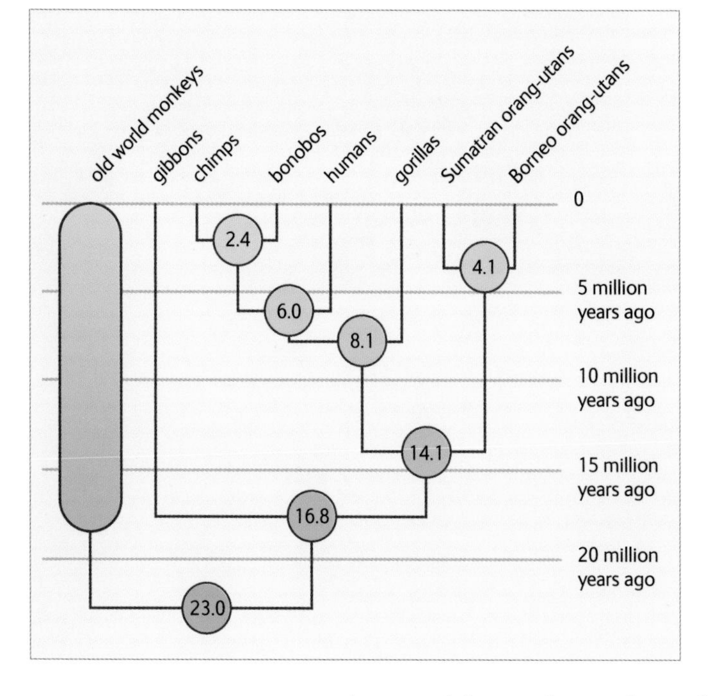

The average rate at which mutations occur is known and provides a powerful technique for estimating evolutionary time, known as the molecular clock. The clock has to be calibrated against a known dated event in the fossil record. Imagine taking mtDNA samples from two humans, analysing the order of the letters in the code, laying the two sequences one above the other and then counting the number of differences. The greater the number of differences, the further away in time was their common ancestor. Knowing the number of differences and the average rate at which changes occur would allow you to estimate when the separation from that common ancestor happened. It sounds easy, but in fact it is a very complicated process and slightly controversial because of assumptions that have to be made. For example, it relies on the average rate at which mutations occur, which is not the same in all regions of the genome, or in all organisms.

Complete sequences of mtDNA have been obtained from living primates and have been used to draw up an evolutionary history that is based on divergence dates – the dates at which a common ancestral line diverged into two branches. The diagram has to be calibrated to the fossil record. The point at which gorillas diverged from the chimp–human lineage is set at 8.1 million years ago by the mtDNA data, but in 2007 a fossil gorilla-like animal was described that is dated to 10.5 million years ago, so some re-calibration is likely to be required.

The divergence between humans and chimps occurs at 6 million years ago, on the basis of the mtDNA evidence. This date accords reasonably well with the dating of fossils that are believed to be close to the common ancestor of chimps and humans. The earliest of these is a striking skull, known colloquially as 'Toumai'. Toumai was dug out of the Sahara desert in Chad by Michel Brunet and Patrick Vignaud and is remarkably complete. The brow ridges, face and relatively small canine teeth are human characters whereas the back of the skull is like that of other fossil apes. The combination of characters is distinct and has led the finders to suggest that this skull is from the earliest known human. It cannot be dated absolutely, but comparison with other fossils in the same strata gives a date of 7.0 to 6.5 million years ago, which fits reasonably well with the mtDNA date for the divergence of humans and chimps.

Toumai is the latest in a long line of fossils to be dubbed 'the earliest human', but unlike the others, Toumai's position is supported by independent molecular evidence. There are reservations among experts, though. The aperture through which the spinal cord passes to reach the brain is further back than in later humans known to walk upright on two feet, and – as habitually walking on two feet distinguishes humans from other apes – it is argued that Toumai must therefore be closer to chimps than humans. Also, a recent re-analysis of the mtDNA divergence date gives a range of 6.3 to 5.4 million years ago. Unless more discoveries are made which change our views, this skull is the closest we can get to looking into the face of our first African human ancestor.

ABOVE Toumai, or *Sahelanthropus tchadensis*, possibly our earliest known human ancestor, was discovered in 2001 in the South Sahara desert.

6 CHAPTER *An eye for Darwin*

CHARLES DARWIN SAID THAT THE THOUGHT OF EYES made him grow cold, so difficult was it to imagine how they might have evolved gradually, with each stage improving upon the last and hence being favoured by natural selection. Modern research has revealed that the process is much easier than he thought, and has happened in much the same way in many different kinds of animals. Master genes, most of them very ancient, determine basic eye structure and where on the body they form, and also control many genes with more minor roles. Eyes illustrate how complex organs can evolve by natural selection, gradually modifying the action of ancient genes that form a wide variety of eyes.

OPPOSITE A great grey owl, *Strix nebulosa*, reveals its yellow irises surrounding large black pupils through which light enters the eyes. These large owls prey on small mammals in boreal forests of Europe and North America.

Most animals are sensitive to light and six out of the 33 major groups, which together account for 96% of all living animals, have eyes with lenses that form images. During the past 20 years, psychologists have made much progress in understanding what different kinds of animals can see and how they use the information. Biologists have compared the molecular and cellular structure of the major components of the eyes of flies, crabs, worms, octopuses and squid with those of fish, chickens, mice and other vertebrates: the lens that focuses the light, the pigments that absorb light energy and the nerves that carry the messages back to the brain. Above all, detailed studies of eyes have taught us how genes make animal bodies.

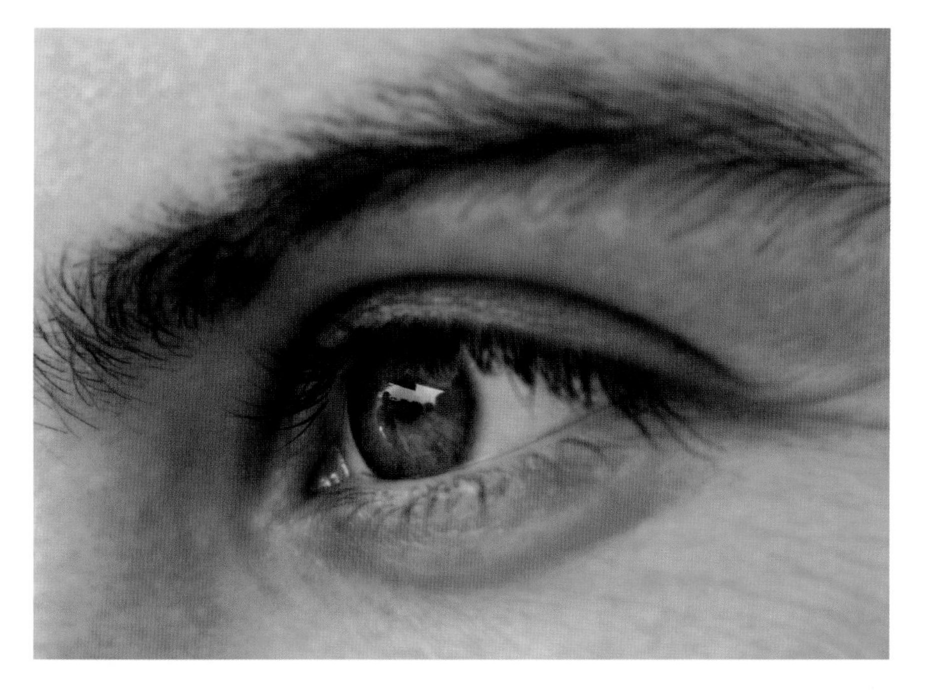

LEFT The human eye is a rotatable globe enclosed in tough tissue, white where exposed, and protected by closable eyelids. Light enters through a circular aperture surrounded by an expandable coloured iris.

EYE DESIGN

Eyes work like cameras: a transparent, curved lens focuses the image on a light-sensitive surface, the retina, which contains pigments that absorb the light and convert it into nerve signals. Like cameras, eyes do not work well if there is too much or too little light; the iris, familiar to us all as a brown, blue or green coloured ring, adjusts the amount of light passing through the central hole (pupil) to the retina behind it. When viewed from the outside, our retina – and that of dogs, horses and most birds – appears black because the pigments it contains absorb a lot, but not all, of the light. The eyes of cats, owls and many other nocturnal animals shine because some of the light is reflected by a mirror-like layer called the tapetum lucidum at the back of the eye. The reflected light has a second chance of being absorbed by the retina, thus improving the sensitivity of night vision. The retina contains millions of physiologically active cells that use lots of energy and hence require a good blood supply.

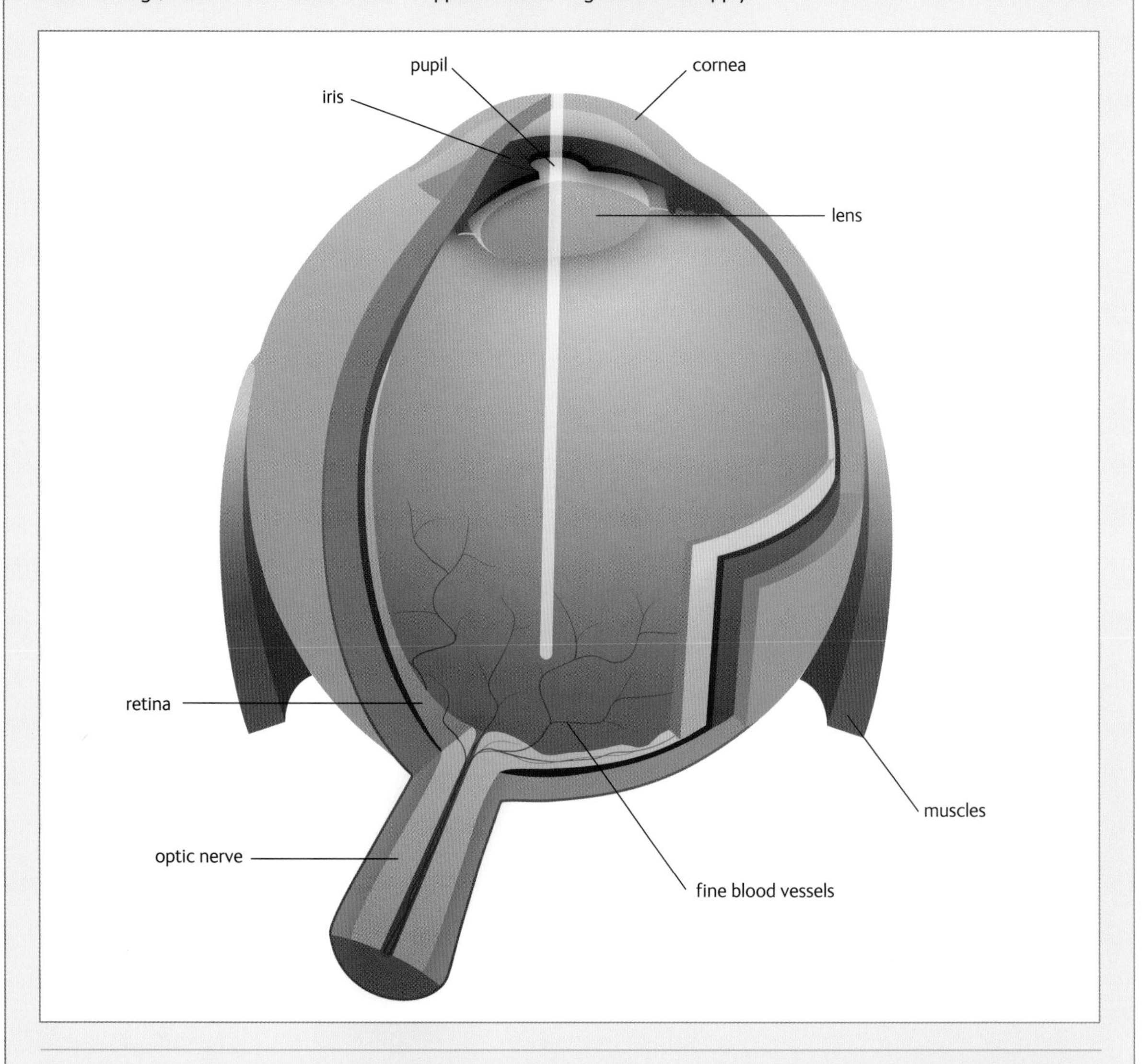

ABOVE Stylized cut-away of a vertebrate eye. Light (white line) enters through the transparent cornea, passes through the adjustable hole at the centre of the iris, through the lens and transparent material in the eyeball to the light-sensitive retina. The millions of tiny, retinal cells are nourished by a network of fine blood vessels that enter the back of the eye with the optic nerve. Muscles swing and rotate the eye.

Like birds and many fish and reptiles (but in contrast to many other mammals), vision is our most important sense. We think of vertebrate eyes as wonderful, but their design is far from ideal. Blood vessels containing light-absorbing red pigments pass in front of the retina, there is always a blind spot in the retina, and the lens and the retina often fail with age. Our eyes are sensitive to only a narrow range of radiation (the visible light spectrum) and work well only in bright light. Most nocturnal mammals rely more on hearing than sight to find food and avoid predators. Many, including almost all bats, have small, almost functionless eyes.

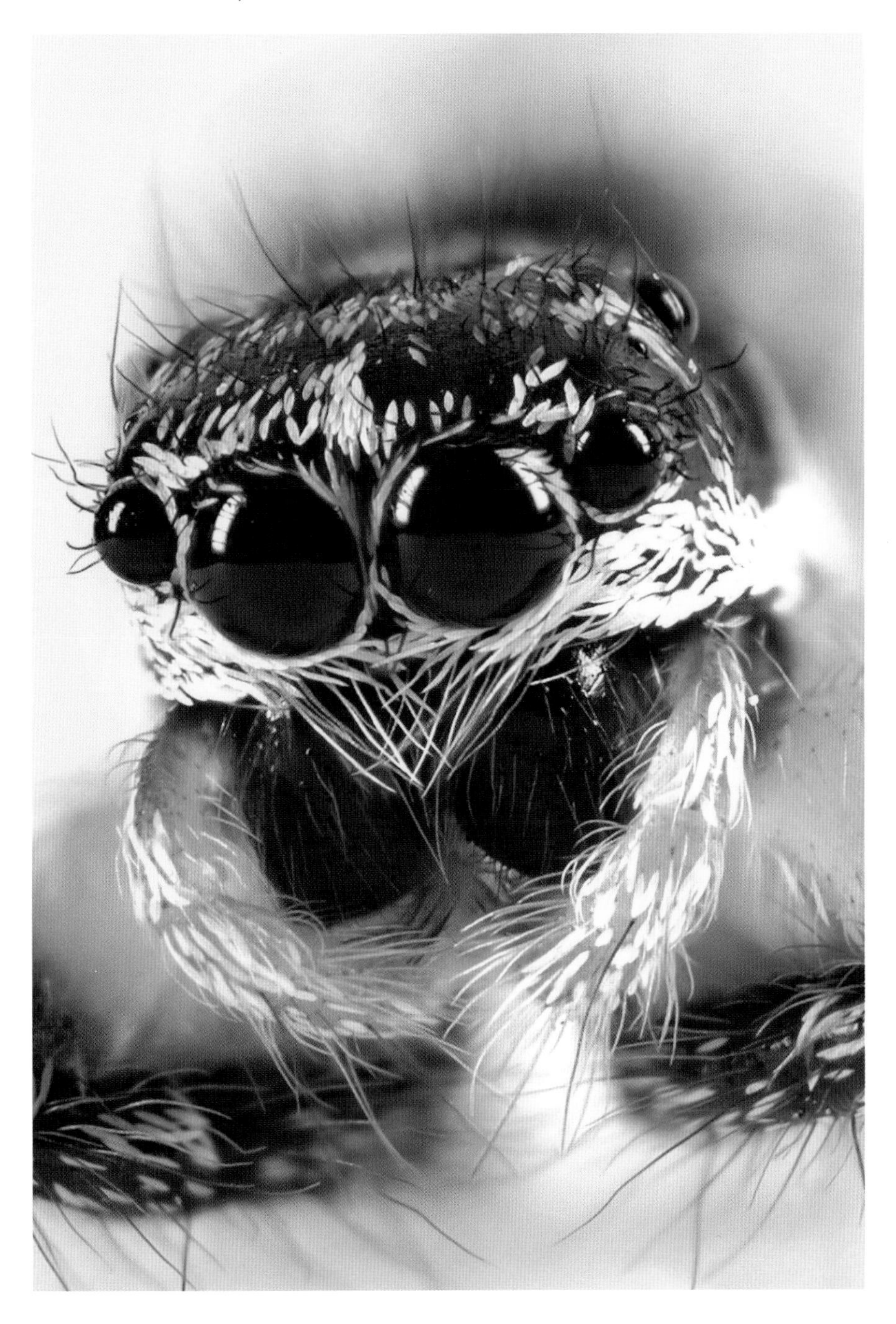

LEFT Most spiders have four pairs of eyes of different sizes and positions according to the species' habit and habitat. Their structure is broadly similar to human eyes, though independently evolved.

LEFT The eyes of squids and octopuses are remarkably similar in structure and composition to those of humans and other backboned animals, although the molluscan body plan and other organs are very different.

Not only can the design of the vertebrate eye be improved upon, but a better design has actually evolved among the molluscs. The eyes of squid and octopuses superficially resemble those of birds, mammals and other vertebrates but are more efficiently constructed: blood vessels pass behind, not in front of, the retina. Many squid and octopuses can distinguish the plane of polarization of light, which gives them additional visual information, and they can see much better in dim light than vertebrates can. Many live in the deep sea where there is very little light, but the huge size of their eyes shows that vision is still an important sense.

Even insect eyes have features that we lack. Most insects see a wider range of colours than we do, are sensitive to the plane of polarization of light and can detect movement more efficiently – which is why it is so difficult to swat flies!

Something old, something new, something borrowed, something blue

The master genes that control eye development and those that specify visual pigments are very ancient and unaltered. They operate in a wide variety of eyes in many different kinds of animals including worms, flies, crabs and molluscs, as well as all other kinds of backboned animals. Visual pigments are the very sophisticated molecules that absorb light energy and, provided they are appropriately located in cell membranes, convert it into nerve signals. The genes that direct the synthesis of the few molecules that work really well in this role seem to have evolved early in the history of life, probably before eyes appeared, since animal visual pigments resemble light-absorbing molecules in bacteria. Since visual pigments apparently cannot be bettered, their genes, and consequently the proteins that they produce, have been retained almost unaltered in otherwise very dissimilar animals.

OPPOSITE Insect eyes are of very different structure to those of spiders and most other kinds of animals. The single pair of large eyes can cover much of the head, and many insects can see extremely well.

In contrast to the invariance of the visual pigments among different animals' eyes, the molecules that form the transparent parts of the eye including the lens differ fundamentally in flies, worms, jellyfish, squid and other animals. The main condition for transparency is that the molecules are aligned in regular arrays, like cars in a car park, so the light can pass through without being scattered or reflected. As long as they are concentrated enough and in a salt solution of appropriate composition, many large biological molecules assemble in this way, so, surprising though it might seem, lenses evolve easily.

Analysis of the composition of eye lenses and their genes in many different kinds of animals shows that various enzymes, structures and even energy production systems have been modified to form transparent materials. Such molecular opportunism in evolution means that, in many cases, the ancestral genes still function in the heart, skeleton, brain or muscles, while the lens-forming genes derived from them are active in the eye. 'Family' relationships between genes of this kind help to explain why congenital diseases of these systems are often also associated with defects of the transparent tissues of the eye.

Eyes use a significant proportion of an animals' energy budget because they contain so many active living cells. The eyes of land-dwelling vertebrates must be kept moist and protected from disease and mechanical damage. Eyelids, blinking and the secretions that in large quantities become tears perform these functions in mammals, birds and many reptiles. Snakes lack eyelids but the eyes are protected by transparent skin that is shed periodically. As Darwin realized, useful vision depends upon the many specialized components of eyes being assembled correctly.

Because eyes have evolved independently in many different kinds of animals, it was assumed until quite recently that each major group of animals had its own molecular toolkit that controlled their embryonic development, but experiments using gene technology have revealed that very different kinds of eyes develop in surprisingly similar ways. Transferring copies of whole genes and groups of genes from one kind of animal to the egg or early embryo of another is one way of finding out what the genes do. In a few strains of the fruit-fly *Drosophila*, the eyes fail to develop; artificially transferring eye-control genes from mice into such fly embryos early in development restores normal eyes. The success of such experiments shows that the genes that form the pigment and all the other elaborate materials of the eye were present in the eyeless flies but were inactive without their 'commanding officer', the control genes; 'commanding officers' from mice work as well as the flies' indigenous ones, even though many of the genes being switched on and the proteins they produce are obviously different. The persistence of genes that produce useful, unique proteins such as visual pigments is easy to understand; why the genes that control eye formation should also be unaltered is far from obvious.

Such experiments were a scientific sensation during the final five years of the 20th century because they revolutionized our understanding of the evolution of genes and the structures that they control. The previous assumption that structures that had evolved independently – fly eyes and mouse eyes, for example – were produced by completely different sets of genes was proved wrong. Biologists now believe that

complex organisms are more like the children's construction toy, Lego: many different models can be assembled from the same set of similar components.

Furthermore, it seemed so amazing that tiny cell nuclei could contain all the instructions for building large bodies that we assumed that most genes were active in most organisms most of the time. In fact, only a small fraction of the genetic material, as little as 5% in some species, actually directs the formation of proteins, though all of it is copied and passed on from generation to generation. To continue with the Lego analogy, we now know that for most constructions, most components and assemblies of components, remain unused in the box, though they were probably used previously and may be used again. This insight enables us to understand how animals with large sophisticated eyes and good vision can evolve easily from ancestors whose eyes were much reduced in size and complexity. As we shall see, the evolution of limbs in our fishy ancestors is another example of the extraordinary capacity of evolution to create novelty from pre-existing structures and genes.

ABOVE Snake eyes are essentially similar to those of all other backboned animals but they lack movable eyelids, so snakes appear to stare.

7 CHAPTER *A fish out of water*

TWO-THIRDS OF THE MODERN EARTH IS COVERED WITH OCEAN; at times in the past, the proportion was even higher. Life originated in water and the first complex animals were aquatic, most of them marine. The majority of major groups have stayed there. Marine animals are very diverse but most of the large, free-swimming species belong to two major groups: fish (vertebrates) and squid, octopuses and cuttlefish (cephalopod molluscs). In spite of being abundant in parts of the ocean, especially the deep sea, and a few species growing to enormous size, cephalopod molluscs have never lived on land (or in freshwater).

OPPOSITE Palaeozoic seas, lakes and rivers teemed with several contrasting lineages of fish that differed in the structure and arrangement of the skeleton, head, jaws, teeth and gills, the fins and the tail.

Fewer than a quarter of the major categories of animals include species that spend their entire lives on land. Only in the arthropods, to which insects and spiders belong, are there more terrestrial than aquatic species. Earthworms, woodlice and land snails are fully terrestrial but most of their relatives, both now and in the past, live in water. The same is true of our own group, the backboned (vertebrate) animals: four-legged (tetrapod) vertebrates, birds, mammals, reptiles and amphibians may be conspicuous to us land dwellers, but both now and throughout evolutionary history there have always been far more species of fish. So, how did our fish ancestors make the transition to land? This question is also about the evolution of how animals' bodies are built, because living on land requires some specialized body features.

LEFT Most modern bony fish have symmetrical tails and fins, including two sets of paired fins, supported by stiff, bony rays moved by muscles inside the body. They are powerful, agile swimmers and most are neutrally buoyant in water.

RIGHT Many different kinds of animals live partly in water and partly on land. Hippopotamuses can stay submerged for several minutes, swimming or walking over river beds.

Becoming terrestrial

Breathing air and possessing legs are two obvious features of land animals, but both can be useful in certain aquatic environments where the transition to land may have begun. Breathing from water passing over gills, as most fish do, works best in clean, well-aerated water, but is almost useless in stagnant, muddy water like that of tropical swamps. Many fish that colonize shallow, slow-flowing water breathe air through lungs as well as, or instead of, breathing through gills. Inflated lungs also contribute buoyancy, making the animal weightless in water, or very nearly so. Several lineages of bony fish, including some of the earliest with jaws, had lungs as well as gills and so presumably breathed air.

Legs, fins, tails and other appendages involved in movement are among the most recognizable features of vertebrate fossils, and so have been intensively studied. Many bottom-dwelling aquatic animals, including lobsters and hippopotamuses, move under water using their legs like paddles or punt poles. So, having legs and breathing air are not by themselves adequate evidence of being terrestrial. The evolution of the ability to walk efficiently on land using legs with articulating wrists or ankles supported by hips and shoulders seems to have been a major challenge.

Fishy relations

Medium-to-large fish first appeared in the Ordovician (495–443 million years ago) and by the Devonian (417–354 million years ago) more than a dozen distinct groups can be recognized, some with jaws and some without, some with bony skeletons and some with skeletons of cartilage (gristle), some freshwater and some marine. Many were powerful predators with massive teeth. Some groups have become extinct; others are represented today by just a few species, among them the small group that evolved into the terrestrial ancestors of tetrapods, amphibians, reptiles, mammals and birds.

Typical modern bony fish are by far the most abundant and diverse group of backboned animals, but the group is relatively recent: it first appeared in the Jurassic (206–142 million years ago) and did not become abundant and diverse until the Cretaceous (142–65 million years ago). We know a great deal about their complicated evolutionary history because not only is the fossil record quite abundant, but also a surprising number of lineages representing intermediate stages in the evolution of modern fish are still not completely extinct: although they have diversified hardly at all through millions of years, a few species survive.

An example is the Mississippi paddlefish, one of only two living species (the other occurring in the Yangtze River in China) of a lineage that first appeared about 400 million years ago. Ancient relatives of these paddlefish evolved to become the ancestors of modern fish and most of their kind died out more than 50 million years ago. Biologists still cannot explain why this large, long-lived fish – paddlefish are among the largest freshwater fish in North America – persists when so many ancient lineages have been entirely replaced by modern kinds of fish. Their closest living relatives are the sturgeons, also large freshwater fish, of which 26 similar species live in major lakes and rivers of the northern hemisphere. One quite small group had fins supported by internal bones and muscles, and hence were named fleshy-finned fish. These fins swivel further on the hip

BELOW Like its extinct relatives, the Mississippi paddlefish, *Polyodon spathula*, has an asymmetrical tail and two pairs of fins on the ventral side of the body. The highly specialized head extends into a large 'paddle' above the huge mouth.

RIGHT The Australian lungfish, *Neoceratodus forsteri*, inhabits rivers and pools in tropical Queensland. It has two pairs of fleshy, lobed fins, a bilobed tail and can breathe with lungs as well as gills, but spends less time on land than other living lungfish.

and shoulder than those of any other bony fish and can bend like toes. Observations on living coelacanths reveal that these fish already had the alternating movements of the fins typical of tetrapod walking.

Many different species, some as large as 3 m (10 ft) long, have been found as fossils throughout the world in sediments deposited during the second half of the Palaeozoic era. The species we know most about lived in tropical freshwaters, but some were certainly marine. Although they were never very abundant, these fleshy-finned fish were clearly well adapted to this habitat because some of their descendants, lungfish,

are still flourishing in very similar environments, 400 million years later. Three somewhat different kinds of lungfish live in tropical west and central Africa, in the Amazon region of South America and in Australia. They feed and breed in lakes, swamps or slow-flowing rivers, using their paired fleshy fins as paddles. The young mostly breathe through gills, but when their ponds and rivers dry up, more mature fish can spend long periods lying dormant encased in mud. Even when active, they can live for many hours out of water, breathing air with their lungs. The adults, especially of the African and American species, breathe air even when in water, surfacing from time to time as whales and seals do.

HIDING IN THE DEEP BLUE SEA

ABOVE A Jurassic coelacanth (*Undina pencillata*) found in southern Germany. This fossil's skull, fins and bilobed tail closely resemble the living *Latimeria*.

Another major group of fleshy-finned fish was first described from a fossil tail in 1839 by the Swiss-born American scientist, Louis Agassiz (1807–1873), who made many fundamental discoveries in fish palaeontology, but rejected Darwin's theories and supported creationism. The tail skeleton consisted of hollow spines, so Agassiz named the group coelacanth (Greek for 'hollow spine'). These large predatory fish were widespread during the Late Palaeozoic, but became rare during the Mesozoic era and were believed to be extinct since the Cretaceous. Just before Christmas in 1938, a trawlerman whose home port was East London on the east side of the South African Cape caught an unusual, large fish. He invited the young curator of the local natural history museum, Marjorie Courtenay-Latimer (1907–2004), to examine it. Fortunately, she had been interested in fossil and living fish since she was a teenager and, although the specimen was already badly decayed and stinking, she recognized its importance. She consulted other experts and together they reached the astounding conclusion that the fish was a coelacanth.

Preserving even small fragments of a 55 kg (120 lb) fish in the heat of the South African summer was a major challenge before refrigeration was readily available. Only fragments of the first coelacanth survive, but the find stimulated others to search for further specimens. Chief among them was fisherman and fish expert, Professor J.L.B. Smith (1897–1968), who named the new species *Latimeria* in honour of Marjorie Courtenay-Latimer and devoted most of the rest of his life to coelacanth research. After 14 frustrating years, a second specimen was caught, again around Christmas, in 1952, in deep water near the Comoros Islands, a remote archipelago surrounded by coral reefs at the north end of the wide channel that separates Madagascar from East Africa, more than 1600 km (1000 miles) from the Cape. Like the first specimen, it was dead by the time scientists reached it.

It was not until 1987 that German scientists using a bathyscape first observed and photographed living coelacanths in their natural surroundings. Like most bony fish, they are weightless in water, and are slow but agile. They paddle in and out of caves backwards, forwards and with the head down, moving their long, mobile fins alternately, as in walking. Although they are predators on other fish, their slow swimming uses energy at a very low rate. Their large eyes seem to be adapted to dim light as they are active only during moonless nights. They do not lay eggs but give birth to live young after a gestation period of many months, may take 20 years to grow to maturity and live as long as a century.

All the finds were in the western Indian Ocean so biologists were again startled when, in 1998, a honeymoon couple noticed a coelacanth in a street market near the north coast of Sulawesi, Indonesia. Another was caught in the same area later the same year and a third in May 2007. Indonesian coelacanths are brown while those from the western Indian Ocean are mottled blue but otherwise they are very similar. Small differences in DNA suggest that they are separate species.

The finding of living coelacanths raises some important questions. Why have no fossils more recent than the Cretaceous been found? Have all the species of that lineage been confined to the deep sea where fossilization is improbable? If so, where? The Comoros Islands themselves are comparatively young, created by undersea volcanoes that first erupted around 10 million years ago and are still active. An attempt at a census around 1990 concluded that the total population of coelacanths around the Comoros Islands was below 650 and probably falling. Are there other species or populations of coelacanths elsewhere? Small silver models that accurately represent the coelacanths' distinctive shape, probably made in Spanish Central America at least two centuries ago, offer intriguing hints that there might be, or at least were until very recently.

ABOVE The coelacanth, *Latimeria chalumnae*, swims slowly using its two pairs of lobed, fleshy fins as well as its bilobed tail.

Extinct and extant amphibians

BELOW Around 365 million years ago (Upper Devonian), *Acanthostega* lived like a large fish-eating salamander, breathing through gills and walking and swimming in freshwater, but also able to emerge onto land. It had a fish-like tail and eight toes on each foot.

As the name implies, typical amphibians (*amphi* means 'both', and *bios* means 'life') spend part of their lives in freshwater (none can tolerate seawater) and part on land, and thus represent transitional stages in the evolution of terrestrial animals from aquatic ancestors. In most amphibians, the eggs and juvenile stages are aquatic, and the adults, which may be very different in structure and habits, are terrestrial, returning to the water only to mate and lay their eggs.

Frogs and toads are by far the most diverse kinds of amphibians that have ever lived. They are not known at all from before the Mesozoic era and did not become abundant until the Tertiary. Ancient amphibians were more similar in body form to the other major living group, newts and salamanders. Although some grew larger than any modern amphibians, fossil amphibians were never very abundant or diverse, but because of their central position in the transition from water to land, they have been studied a lot. Much of what we know comes from fossils found in sites in Greenland and northern Canada where, during the Late Devonian about 360 million years ago, there were equatorial meandering rivers. Fragments of more than a dozen specimens collected in northeast Greenland in 1931, and later finds in Russia, Australia and China, revealed many similarities between ancient amphibians and fleshy-finned fish resembling lungfish. Although bones corresponding to the arms and legs can be identified, some of these early amphibians had 6–8 fingers or toes on each limb, suggesting that the organization of the long bones was established some time before tetrapods 'made up their mind' about the structure in which hands and feet have five digits.

BELOW The great crested newt, *Triturus cristatus*, breeds in freshwater but adults spend long periods on land. These amphibians can breathe air and have four jointed limbs with five or fewer toes and a long, flexible tail.

ABOVE *Tiktaalik roseae*, an air-breathing amphibian with skull, neck and ribs similar to those of early tetrapods and fish-like jaw, fins and scales. The rocks in which the fossil was found were formed in shallow tropical rivers and lagoons.

Fish fingers

None of these fossils fulfils all expectations for transitional species between fish and amphibians. The breakthrough came with the discovery in 2004 of three fossils, one an almost complete skull, of a hitherto unknown species that had the mix of characters expected of an intermediate between fish and tetrapods. At the suggestion of the local Inuit on Ellesmere Island, northern Canada, where it was found in rocks of Late Devonian age, palaeontologists named the fossil *Tiktaalik*. Its front limbs emerged from a stout shoulder girdle and had flexible wrists, suitable for supporting the body weight in air, though they lacked proper hands and fingers. Although its skull was flat with upward-pointing eyes like aquatic frogs, *Tiktaalik* had a distinct neck that enabled the head to move relative to the shoulders and the snout to tilt downwards, essential for eating food on land. These characters suggest that *Tiktaalik* was a true amphibian, able to live on land, but its young were probably aquatic and it frequently walked or swam in shallow water, as modern crocodiles do.

All Carboniferous and later tetrapods have five (or fewer) toes; palaeontologists believe that all are descended from one (possibly two) ancestral species that successfully colonized land. The other Devonian amphibians with larger numbers of fingers and toes died out without descendants. Reduction of digits is very common: bird feet never have more than four, and living hoofed mammals only one or two. Although mutant animals (and humans) with extra digits are quite common and can easily be induced experimentally, very few tetrapods have more than five digits. Only ichthyosaurs - Mesozoic marine reptiles that resemble whales in having fins, a swimming tail and giving birth at sea - had flipper-like forelimbs that clearly deviate from the typical plan.

Resurrecting genes

As in the evolution of eye components, ancient genes can perform pivotal roles in the formation of novel structures after millions of years in obscurity or inactivity. To explore the changes in gene action that might have produced early hands and feet, scientists studied the development of fins and limbs in the early embryos of fish, mice and poultry. Unfortunately, they chose the zebrafish, which is easy to breed and study in the laboratory, but it is an advanced, modern bony fish that is very distantly related to tetrapod ancestors, and has lost (or never had) some key genes.

Both fins and tetrapod limbs begin as paired ridges, but the development of tetrapod limbs has a second stage involving other genes that direct the formation of discrete arms and legs. These additional genes were 'new' to tetrapods, perhaps appearing first in *Tiktaalik* or its relations. Further studies showed that these same genes are active in embryos of the Mississippi paddlefish where they control the formation of the paired,

BELOW Tree showing the relationships between the living and fossil animals.

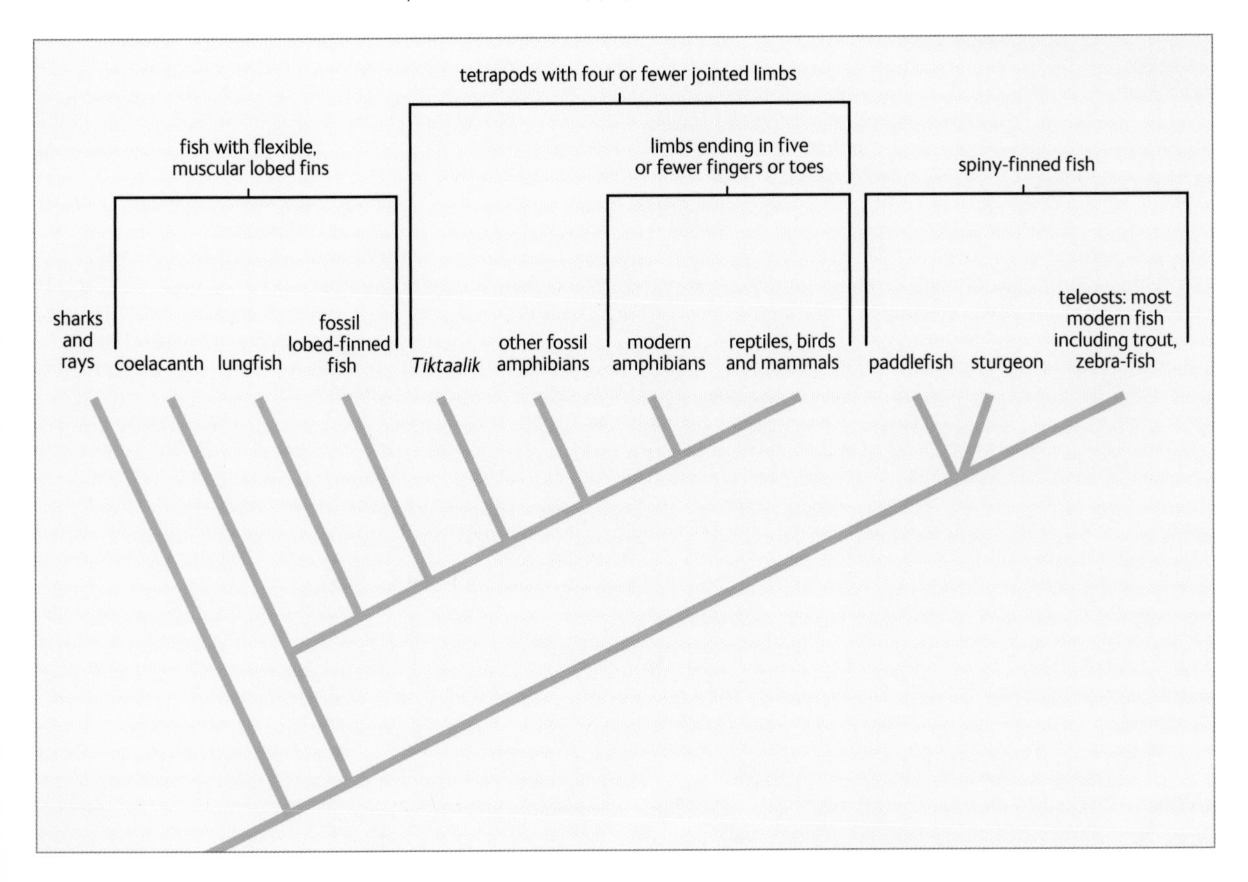

paddle-like fins. Genes found in distantly related modern groups were presumably present in their common ancestors. So *Tiktaalik* probably already had genes able to form paired limbs; they are functional in amphibians and other tetrapods but silenced or absent in most modern fish. Are these control genes also active in coelacanths, perhaps helping to form the fleshy fins? This question awaits the opportunity to study coelacanth embryos in the laboratory.

Many were called, few were chosen

Very few lineages of animals have become completely terrestrial, though many have some features, such as limbs and air breathing, that would equip them for life on land. Stout limbs attached to girdles and ending in forward-pointing hands or feet are adaptations to walking on land, but having no more than five fingers or toes may be accidental; this feature was present in the direct ancestors of tetrapods and has been retained ever since.

The contributions of such 'neutral' or 'adaptive' characters to the evolution of body forms are hotly debated. The ancestry of key genes in the formation of limbs can be identified because a few species of a surprising number of lineages of fish are still extant. Coelacanths remained unnoticed by scientists until 70 years ago. How many more groups of organisms thought to be extinct survive in the deep sea, remote mountains, under polar ice or in other inaccessible places? As yet undiscovered species may reveal relationships between major groups with contrasting structures and habitats. Could there be anything as yet unknown that is as unlikely sounding as a group of tetrapods that have lost some of their limbs and returned to the sea?

8 CHAPTER *A whale of a problem*

OPPOSITE In most great whales the teeth are replaced by fringes of stiff baleen, around the greatly enlarged mouth and tongue, that strain food from the water. Cetaceans are the most abundant and diverse living group of wholly aquatic mammals.

GIVING BIRTH TO LIVE YOUNG AND SUCKLING THEM ON MILK were sufficient to convince the Greek philosopher and scientist Aristotle (384–322 BC), that whales, dolphins and porpoises were mammals. Whether they were primitive mammals that arose directly from aquatic reptiles or even fish and were the ancestors of typical mammals, or had arisen from terrestrial mammals that had secondarily gone back to the water, was much discussed during the 18th and 19th centuries. Darwin himself favoured the latter scenario and in the first edition of *On the Origin of Species*, he suggested that whales evolved from bear-like ancestors that swam with the mouth open, collecting floating insects. For terrestrial mammals to become marine may seem improbable to us, but in fact the fossil record suggests that it has happened at least seven times.

After the dinosaurs

The end of the Cretaceous is defined by the extinction of many animals, most famously the dinosaurs. Mammals and birds were among the groups that did not become extinct; some – the fossil record is vague about exactly which – survived the mass extinction. Mammals evolved rapidly during the Early Cenozoic, producing representatives of nearly all the modern groups, including bats, primates, rodents (rats, mice, squirrels, guinea pigs), carnivores and both major groups of hoofed, herbivorous mammals, (horses, tapirs and rhinoceroses), and artiodactyls (pigs, hippopotamuses, deer, camels, llamas, giraffes, antelopes, cattle, goats and sheep).

LEFT Otters are semi-aquatic, with short limbs ending in webbed feet, muscular bodies, reduced ears, closable nostrils and water-resistant fur. They are agile hunters, catching fish and other prey in shallow water.

Most Cenozoic mammals were terrestrial, as their Mesozoic predecessors had been, but several groups became partially or entirely aquatic. Some are now extinct, but extant semi-aquatic mammals include otters, which are closely related to stoats and weasels, and like them are active hunters, but they catch fish under water, instead of rabbits and voles on land. Beavers, muskrats, water rats and voles, capybaras and coypus are freshwater rodents that eat terrestrial as well as aquatic vegetation and shelter in and near lakes and rivers. All these semi-aquatic mammals have webbed feet and other adaptations for swimming, and can dive a few metres down for a few minutes at a time, but in other ways are similar to their terrestrial relatives.

ABOVE Sea-lions are thoroughly specialised for marine life, with greatly reduced ears and tail, short, dense fur, short limbs with relatively large hands and feet adapted as flippers, but they breathe air and breed on beaches.

Three living groups of mammals are more completely committed to aquatic life and are predominantly marine: pinnipeds (seals, sea lions and walruses), sirenians (manatees and dugongs) and the cetaceans (whales, dolphins and porpoises). Seals and their relations live mainly near coasts and in estuaries where they dive tens of metres (yards) underwater, chasing fish, seabirds and other prey. Their graceful, agile swimming is powered by up-and-down movements of the body and all four legs, which are modified as flippers, but pinnipeds can also move, hear and see quite well on land. They all mate, give birth and suckle their pups out of water, usually on beaches or, in the case of some Arctic species, on ice floes.

Aquatic for life

The other two groups of marine mammals pass their entire life-cycle in water. They never come out onto land and soon die if stranded. Their limbs, eyes, ears and jaws function well only in water.

Manatees and dugongs are herbivores, eating seaweeds and other aquatic vegetation, mostly in sheltered coastal waters, mangrove swamps and estuaries. They swim under water but cannot dive deeply.

Cetaceans are primarily oceanic. Dolphins, porpoises and toothed whales are carnivores, eating fish, squid, seabirds and sometimes each other. In most of the very large whales, teeth are replaced by baleen, a substance similar to hair that, with the huge tongue, acts as a sieve, straining out organisms from seawater. Whales are the most proficient divers of all vertebrates. Obtaining accurate information about their performance is difficult, but sperm whales are known to dive thousands of metres down and stay submerged for up to two hours.

WHALE CHARACTERS

Although they live with and in many ways are like fish, all aquatic mammals retain their essentially mammalian characters. All give birth to live young and suckle them on milk. All breathe air through lungs, not gills. All are warm-blooded, and have fur or structures derived from fur-like whiskers. The backbone always bends up and down, not from side to side, as in fish, amphibians and reptiles.

Hearing is the dominant sense in all cetaceans. The special qualities of whale ears are only revealed by internal investigation because the visible, external parts have disappeared: protruding ears would spoil the streamlining of the head so essential for fast, efficient swimming. The parts of the ear concerned with balance are specialized for acrobatic swimming in a variety of postures, and for deep diving. The hearing organ is sensitive to a wide range of sounds that are focused, not by an external ear, but by a fatty mass on the skull. Cetaceans, especially porpoises and dolphins, communicate with each other by elaborate sounds that scientists are only beginning to understand, and most toothed whales navigate by echolocation, emitting high-pitched clicks and identifying objects from their echoes.

Whales swim by powerful up-and-down movements of the tail, which has broad, stout, horizontal flukes made mainly of cartilage, not bone. Some cetaceans, especially fast, agile predators, have fins of similar structure along the midline of the body. The hindlimbs are reduced to tiny lumps of bone and the forelimbs have become flippers, with proportionately short arms and greatly enlarged fingers; their movements aid steering during swimming and are used by mothers to guide their babies. The numerous ribs are not firmly attached to the spine, thus enabling the lungs in the chest to expand and collapse more than in terrestrial mammals, an essential adaptation to deep diving. The head, especially the mouth, is relatively very large, more than a third of the total body length in some species, and the skull is very specialized, with some bones much enlarged while others are reduced. The nose – its popular name, the blowhole, reflects the force with which the lungs can expel gases after a long period of submergence – is on top of the head and far back, almost behind the hinge of the enlarged jaws that form the huge mouth.

ABOVE When they surface to breathe, whales expel stale air very forcefully.

ABOVE Hippos breathe air and feed on land, but adults and young spend the day in rivers; the water dissipates body heat and protects the hairless skin from drying.

ABOVE Hippos graze at night, consuming huge quantities of grass and scrub vegetation. They can run surprisingly fast on their splayed, 4-toed feet.

There is one semi-aquatic mammal that we have not mentioned yet although it is the largest and in many ways the most familiar of all: hippopotamuses – the name is derived from Greek words meaning 'horse' and 'river' – spend much of their time in water but obtain almost all their food on land. Hippos emerge at night to graze on grasses and other terrestrial vegetation, spending most of the day in shallow lakes and slow-flowing rivers, where they digest their enormous meals and evacuate the waste. They are usually seen in herds wallowing with the head and back breaking the surface, but they can swim submerged for a few minutes at a time; the exact motion was revealed only recently by modern underwater photography. Thus, they seem to be about as committed to aquatic life as otters, beavers or water voles. A crucial difference from other partially aquatic mammals is that hippos can give birth and suckle under water. As well as swimming as soon as it can walk, the baby hippo, like newborn cetaceans and sirenians, masters at birth what seems to us the difficult task of surfacing intermittently to breathe. Sharing African rivers with large crocodiles is no problem: hippos are devoted mothers and do not hesitate to attack anything that approaches their young.

Since the middle of the 19th century, standard classifications placed hippos among the Artiodactyla (defined as having an even number of toes: *artio* means 'straight' or 'even', and *dactyl* means 'toes'), alongside the peccaries, pigs, warthogs and their relations. These non-ruminants are omnivores, burrowing for roots and fungi and eating carrion, small prey and nuts with only small quantities of vegetation. Their large litters are born in a burrow or nest. Like hippos, they are stockily built with sparse, coarse hair and lack horns or antlers. The adults' teeth form tusks and what they lack in size, they more than make up for in ferocity.

Hippos are committed herbivores, grazing on grass and herbs, with a huge stomach and long gut of complicated structure, but hippos do not chew the cud as ruminants (such as deer, giraffes, cattle and sheep) do. The stomach is not divided into several chambers and the typical side-to-side movement of the jaws would be impossible with those enormous tusks, which are displayed frequently by opening the mouth unusually wide. But, as in advanced artiodactyls, a single offspring is born without a nest. Hippos are such massive animals that it is not surprising that four splayed toes on each foot support the body weight. The feet of smaller and more lightly built artiodactyls have just two functional toes forming the characteristic cloven hoof. The family Hippopotamidae always sat somewhat awkwardly within the artiodactyls but since there are only two living species – the common hippo (*Hippopotamus amphibius*) found over much of Central and East Africa and the pygmy hippo (*Hexaprotodon liberiensis*) of West Africa – taxonomists were happy to leave them where they were.

ABOVE Pigs and peccaries are primitive, non-ruminant artiodactyls that give birth to larger litters in a nest or burrow. Most are forest dwellers, eating nuts, roots and small prey.

LEFT Goats, along with antelopes, cattle, sheep and deer, are ruminant artiodactyls with cloven hooves. They are the most abundant and diverse grazing and browsing mammals in the modern fauna. The single young follows the herd from birth.

Whale origins

Thirteen months after Darwin's death, on 23 May 1883, the President of the Zoological Society of London, Professor William H. Flower (1831–1899), gave a lecture on whale origins to the Royal Institution in London. He quoted reports of observations that foetal and neonatal whales had hair and differentiated teeth as proving that cetaceans are descended from terrestrial mammals. However, he reasoned, whales could not be derived from seals, walruses and sea lions because reduction of the hindlegs was very unlikely once they had become important for swimming. The powerful cetacean tail would not have evolved from the almost tailless pinnipeds to replace the hindlegs. Earlier comparative anatomists had studied the mouths, guts and reproductive organs of stranded whales and noted their resemblance to those of non-ruminant artiodactyls. From such evidence, Flower proposed that these partially limbless, mainly carnivorous mammals evolved from omnivorous, pig-like artiodactyls.

For more than a century, this idea attracted few supporters. Palaeontologists studying only skeletons preferred the theory that cetaceans arose from a long-extinct group of dog-sized predatory mammals, though they were unable to say exactly how or when. But in the late 1980s, biochemists studying the molecular structure of various proteins and DNA sequences, including mitochondrial DNA and SINEs (see box), from scores of wild mammals belonging to various groups, identified hippopotamuses as the nearest living relative to cetaceans. Their data suggest that, at least 55 million years ago, the early ancestors of typical artiodactyls split off from animals that later evolved into whales and hippopotamuses.

Similar data exclude any close relationship between the pig and peccary families and hippos. The ancestors of the pig and peccary families branched off from those that went on to form hippos and whales at the very beginning of the Tertiary, just after the groups that became ruminants. The many resemblances in body form, tusks, sparse hair and pugnacious temperament must have evolved convergently. The classification clearly had to be revised to accommodate the new information. A new group the Cetartiodactyla, which unites Cetacea and Artiodactyla has been proposed, but has not yet received universal support. The example shows how new laboratory techniques and new fossil finds can revolutionize our understanding of the course of evolution.

OPPOSITE Sir William Henry Flower served as a surgeon in the Crimean War (1853-6) and later directed anatomical museums in Glasgow and London. As well as much original research, he greatly improved museum exhibits and thrilled audiences with his public lectures.

SINES POINT THE WAY

SINEs (**S**hort **I**nterspersed **N**uclear **E**lements) form part of what used to be known as 'junk DNA', genetic material that is not transcribed into RNA or proteins. SINEs are identified by characteristic sequences and are abundant in mammals, representing about 13% of all nuclear DNA. They are small, less than 500 base pairs long, and are scattered around the rest of the genome, often in numerous copies. Since they are functionless, the abundance, composition and position in the genome of SINEs are not altered by natural selection and hence their accumulated random mutations can indicate phylogenetic relationships.

STEPS TO LEGLESSNESS

ABOVE All whales and dolphins breathe air through dorsal nostrils ('blowhole'), have smooth, hairless skin and swim by up-and-down movements of the greatly expanded tail, powered by huge body muscles. External hindlimbs are absent and the forelimbs form flippers.

Since 1975, all cetaceans have been protected species, so collecting embryos and newborns is almost impossible. Fortunately, museums keep specimens obtained through accidental deaths and were able to provide developmental biologists with several pickled embryos of one of the smallest and most widespread cetaceans, the spotted dolphin, *Stenella attenuata*. Biochemical methods of identifying gene action are now so advanced that some can be identified even in preserved material. Such methods were combined with more traditional techniques to show that, early in development, the paired buds forming the hips and legs grow as they would in terrestrial mammals but the genes for the signal molecules driving such growth switch off by about the fifth week of gestation. The rest of the body, especially the tail and body muscles that move it, continues to grow, engulfing the tiny limb buds, which are invisible and functionless by the time the baby dolphin is born 10–11 months later.

Matching these findings to recently discovered fossils suggests that *Pakicetus* and its close relations had fully functional though relatively short hips, legs and feet but, as swimming with the back and tail became dominant, the hindlimbs reduced, starting with the feet. The hindlimbs of modern cetaceans are usually represented only by irregularly shaped remnants of the hip and leg that are tiny compared

BELOW Occasionally whalers find small bones buried in the body muscles of adult cetaceans. They are the functionless remnants of the hip bones.

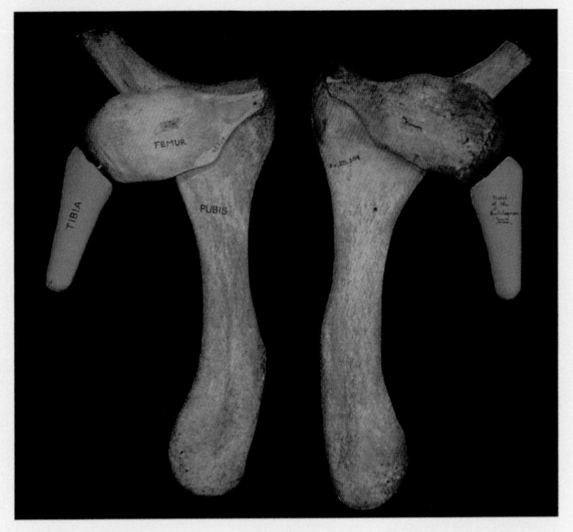

to the strong vertebrae. However, the genes that form hindlegs are still present because very occasionally they remain active in the embryo for long enough to produce vestigial hindlegs. Several million whales were caught over two centuries of industrialized whaling (roughly 1760–1960), among which at least a dozen specimens had noticeable protruding hindlegs.

The loss of one or both pairs of limbs has happened several times among tetrapod vertebrates: as well as cetaceans and sirenians that lack hindlimbs, the modern fauna includes limbless amphibians (caecilians), limbless lizards (and several lineages with much reduced legs), snakes and kiwis (birds whose forelimbs have almost disappeared completely). Detailed studies of python embryos revealed striking similarities in the pattern of gene inactivation during development, suggesting that limblessness evolved by similar mechanisms in snakes and whales, and probably also in the other groups of limbless vertebrates. Vestigial hindlimbs are present in primitive snakes such as pythons and boas, and occasionally in other snakes.

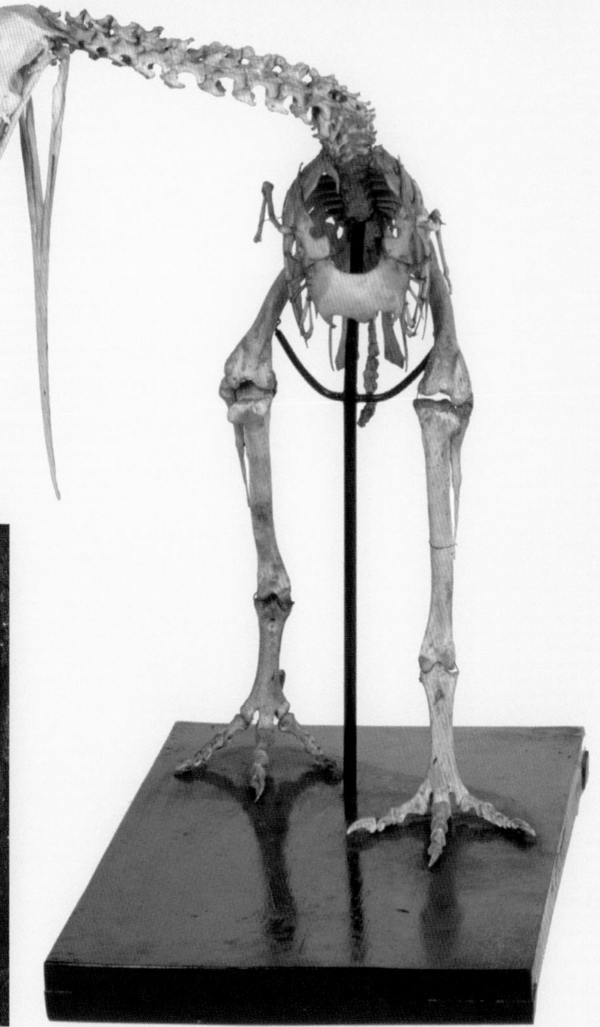

RIGHT Limblessness has evolved many times among tetrapod vertebrates. The most abundant and diverse are snakes (top). Certain burrowing amphibians are blind and limbless (below). Flying birds have relatively huge forelimbs, but the wing bones are tiny and functionless in flightless kiwis (right).

Ancient whales

ABOVE Skull of *Pakicetus inachus*. Its numerous sharp, conical teeth and pointed snout indicate fish eating, but since its nostrils are anterior, not on top of the head as in modern whales, this dog-sized animal probably did not dive.

Until the 1980s, no known fossils threw any light on the early evolution of whales. One of the first to be recognized as a whale ancestor was a skull dating from the Early Eocene found in the Himalayas of northern Pakistan in rocks formed at the edge of the ancient Tethys Ocean. Its discovery prompted palaeontologists to search for more fossils in the same area, leading to the finding in 2001 of a more complete skeleton of this animal, named *Pakicetus*, and several slightly later relatives. These whale ancestors were about the size and shape of a wolf, and had broad, flat heads with the nose in the typical position on the front tip of the skull, suggesting that they may have ambushed prey while partly submerged, as crocodiles do. Their legs, hips and shoulders were strong enough to walk on land as well as wade in shallow water and swim, like modern hippopotamuses.

RIGHT Artist's impression of *Pakicetus*. With four paddle-like limbs and a relatively short tail, this early ancestor of whales probably both swam in shallow water and walked on land.

The distinctive arrangement of the bones around the ear link *Pakicetus* to whales and the bones of the hindlegs support its affinity to artiodactyls. Plant-eating animals, especially grazers, are obliged to spend long periods out in the open where they are exposed to predators. Artiodactyls escape by running far and fast on their long, thin legs ending in firm hooves. The risk of disasters such as twisting an ankle on hard, uneven ground is reduced by sockets in the ankle bone that keep the joint straight. This distinctive feature is also found in the fossils of early whale ancestors; perhaps twist-resistant ankles enabled these amphibious predators to keep their footing on slippery Tethys shores. Whether they suckled their young under water, as hippos do, and whether they could manage without drinking freshwater, cannot be determined from the fossils.

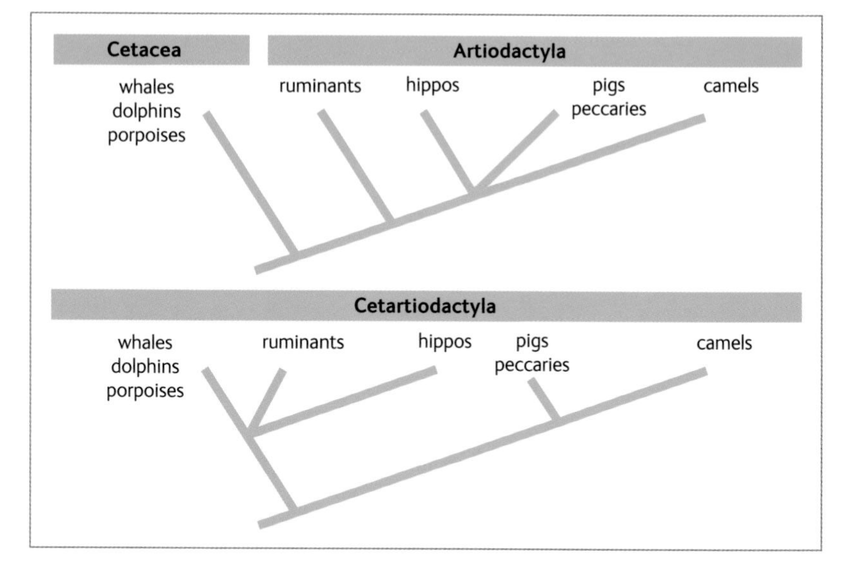

ABOVE Trees showing the long accepted (above) and recently proposed (below) relationships between cetaceans and certain groups of artiodactyls.

Later fossil whales are more completely specialized for swimming at the expense of proficient walking, with the hindlegs and hips reduced in size and not properly attached to the spine, and the tail becoming longer and stronger. By the Late Eocene, cetacean fossils are found throughout the world, suggesting that they had evolved to become completely oceanic during a comparatively short period.

So the modern view of the origin of whales is summarized with uncanny precision by Professor Flower's scenario presented in his lecture to the Royal Institution:

> *We may conclude by picturing to ourselves some primitive generalized, marsh-haunting animals with scanty covering of hair like the modern hippopotamus, but with broad swimming tails and short limbs, omnivorous in their mode of feeding, probably combining water plants with mussels, worms, and freshwater crustaceans, gradually becoming more and more adapted to fill the void place ready for them on the aquatic side of the borderland on which they dwelt, and so by degrees being modified into dolphin-like creatures inhabiting lakes and rivers, and ultimately finding their way into the oceans.*
>
> Flower, W.H. (1883)

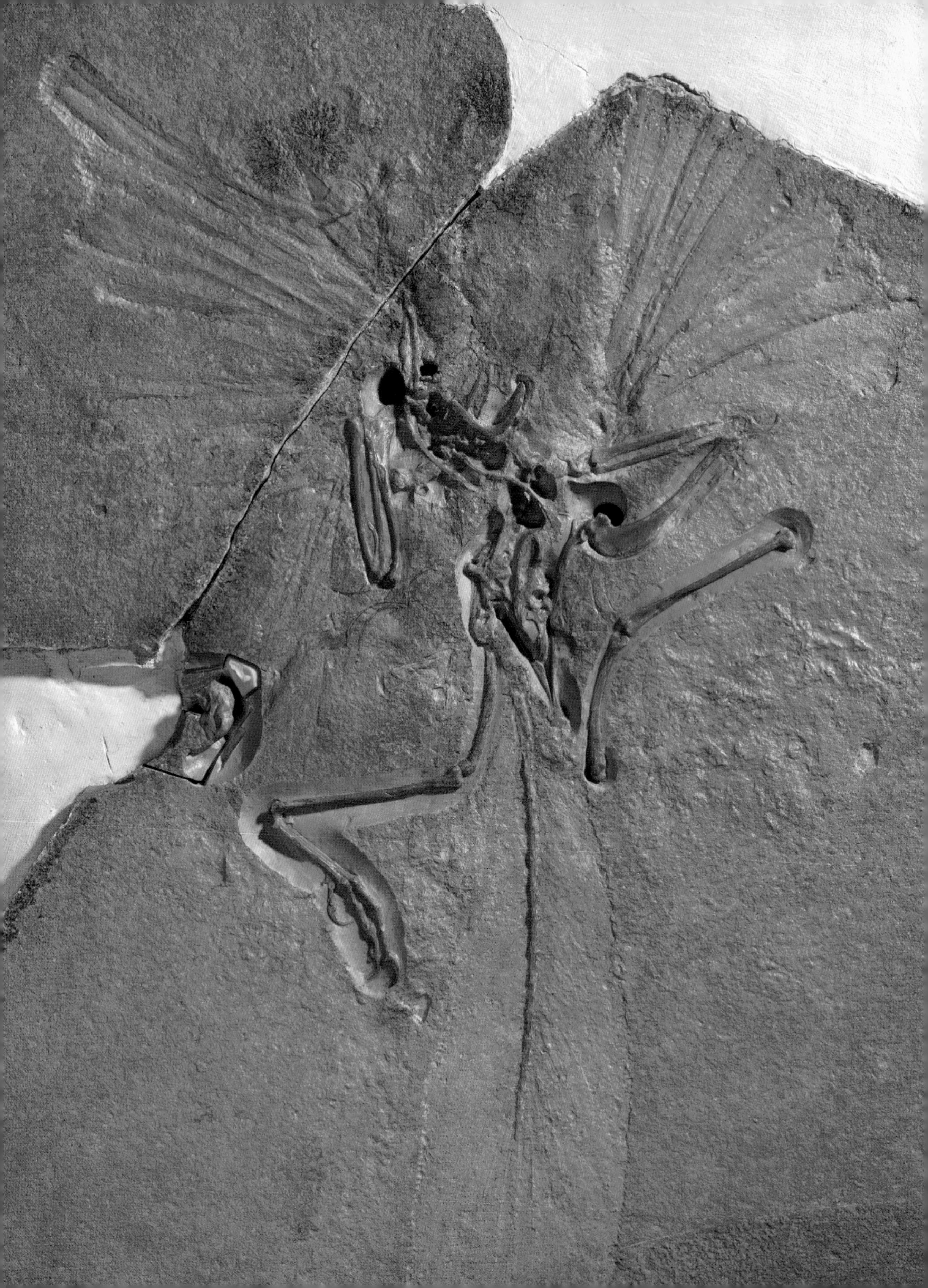

CHAPTER 9 Feathered fossils

THE MID 19TH CENTURY WAS THE HEYDAY of the study of large fossils. Surface sedimentary rocks were cut for building canals and later railways, and steam engines and power tools enabled more and deeper mines. Among the most spectacular discoveries were dinosaurs, first defined and named in 1842. Although only a few species were gigantic, these large reptiles caught the public imagination. Enthusiasm for them was such that, where genuine fossilized bones were unavailable, museum directors displayed plaster casts of the skeletons, mounted in lifelike poses. No one realized that not all dinosaurs were extinct.

OPPOSITE *Archaeopteryx lithographica* fossil in fine calcareous limestone. All known specimens come from a single quarry in southern Germany.

Fossilized families

Britain and northern Europe have few indigenous reptiles, all so small and inconspicuous that they were little studied, but scientists and explorers were reporting the structure and habits of crocodiles, giant turtles and other large reptiles in India, Australia, Africa and South America. Tortoises, terrapins, alligators, crocodiles and most snakes and lizards lay eggs, which closely resemble birds' eggs in structure and in chemical composition. The mother chooses warm places to lay her eggs and may guard the nest until the young hatch. Even so, palaeontologists were reluctant to accept that the breeding habits of at least the more advanced dinosaurs closely resembled those of birds. In 1923, a fossilized nest containing at least six eggs was found with the skull of a small, hitherto unknown, dinosaur only 10 cm (4 in) away. The obvious interpretation – that the dinosaur was a parent guarding the nest – was so successfully challenged by other palaeontologists, who believed that the skull was from a predator, that the new species was named *Oviraptor*, which means 'egg thief'. Since then several different kinds of fossilized nest, some containing more than two dozen eggs, with remains recognizable as embryonic dinosaurs have been found.

ABOVE Reptile eggs are almost identical in basic structure to those of birds. The white spherical egg was laid by the tortoise *Geochelone denticulata*; the brown one is a domestic hen's egg.

LEFT Reptile eggs, like birds eggs, cannot live without air. Marine turtles return to land only to dig nests and lay their eggs.

The similarities are strong evidence for close evolutionary relationships but there are many glaring contrasts between the adult anatomy and physiology of living birds and reptiles. Birds have warm bodies insulated with feathers that are also essential to flight and to many aspects of their extensive, elaborate family life; reptiles have variable, usually low, body temperature, scaly or shelled skin and mostly solitary lives with simple, transient interactions between the sexes and between parents and offspring. Palaeontologists concluded that, since the adults are so different, even though their eggs and breeding habits are so similar, there must be several, probably many, intermediates between reptiles and birds that had not yet been discovered as fossils.

OPPOSITE Almost all birds guard and incubate their eggs in a specially constructed nest, and most feed their chicks.

ABOVE Female alligators guard the nest they have built for their eggs, carry or escort the hatchlings to water and chaperone them for several weeks, usually without feeding themselves.

German aviation and Chinese dragons

In 1861, shortly after the publication of *On the Origin of Species*, the finding of the predicted 'link' between reptiles and birds was announced and named *Archaeopteryx lithographica*. The fossil was preserved in fine calcareous mud dating from the Late Jurassic (about 150 million years ago). *Archaeopteryx* was small, about the size of a crow, with delicate bones, which would not have fossilized readily. The fossils have clear impressions of feathers that are almost identical to those of modern birds, but *Archaeopteryx* probably flew only weakly because it lacked the bones associated with the large, powerful breast muscles that flap the wings in flying birds. It also had numerous teeth but no beak, and a long, reptile-like tail.

In broad terms, the discovery confirmed Darwin's theory of gradual evolution, and although Darwin himself was not impressed (he did not mention the fossil in later revised editions of *On the Origin of Species*), some of his supporters, notably Thomas Henry Huxley (1825–1895), championed the position of *Archaeopteryx* as 'the' intermediate between birds and dinosaurs. His interpretation was much disputed and could not be resolved without further information. New finds and new interpretations clarified understanding of the evolution of amphibians and mammals, but few shed light on the origin of birds. So the exact course of evolution of birds from reptiles, particularly how and when feathers evolved, remained unclear for more than a century.

Around the millennium, two quite separate advances in science again turned the evolution of birds into a hot topic. Firstly, some exceptionally well-preserved fossils of several kinds of hitherto unknown small, bird-like dinosaurs from Yixian, an area of northern China, became

LEFT Crow-sized *Archaeopteryx* had bird-like feathers and its forelimbs formed wings, but it had numerous teeth like its reptilian ancestors and a much longer tail than modern birds.

RIGHT *Microraptor gui* an early Cretaceous dinosaur with wings and feathers found in Liaoning, northern China. This cat-sized predator probably glided rather than flew.

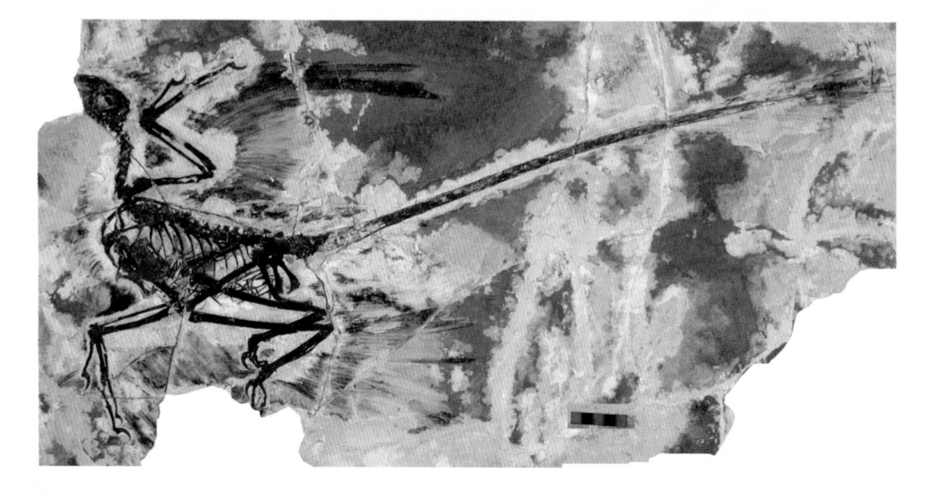

available to international scientists. In the Early Cretaceous, about 125 million years ago, this region had lush forests and lakes near some very active volcanoes that periodically covered the region with ash. Many of the specimens were fossilized in lake sediments but seem to have been fully terrestrial. Most were small, between the sizes of a duck and a small dog. Small animals are less likely to fossilize successfully than large ones and their fossils are less likely to be found, which perhaps explains why very few similar dinosaurs are known from elsewhere in the world.

The Yixian dinosaurs were lightly built, fast-moving predators, some running on their hindlegs like mythical dragons, perhaps catching insects, small mammals and other prey in their grasping forelimbs. Most fossils were fringed with strands that most, but not all, experts interpret as simple feathers that resemble hair and may have shared its function of thermal insulation, perhaps keeping these small animals warm during cool nights. Being so small, the dinosaurs, and especially their babies, would cool and become sluggish during long nights, making them vulnerable to surprise attack from larger predators. A warm coat of downy feathers would have been just the thing for keeping alert and active.

These fossils confirmed what some palaeontologists had long believed: that the evolution of feathers long preceded the emergence of true birds and was not specifically associated with flight.

The structure of the bones suggests that some of these small dinosaurs were ancestral to the well-known Late Cretaceous giant dinosaurs, including *Tyrannosaurus rex*. Keeping warm is even more important for fast-growing hatchlings. Although fossilized impressions of its scaly skin showed clearly that adult *Tyrannosaurus* did not have feathers – its huge body generated more than enough heat – the relatively very small hatchlings may have been covered with down feathers, much as baby elephants are hairy while the adults are almost hairless.

ABOVE *Mei long* 'sleeping dragon', a small feathered dinosaur that was particularly well preserved in fine volcanic ash. Although not a direct ancestor of birds, it had many bird-like features, as suggested by the reconstruction (above).

Chicken feathers

Meanwhile in the United States and Europe, scientists were studying the early development of different kinds of feathers in ordinary domesticated hens' eggs. Most modern birds have just three main kinds of feathers: fluffy down consisting of tufts of strands; flat, stiff-vaned contour feathers that cover the body and most of the head, and even stiffer flight feathers with asymmetrical vanes on the wings and tail. In poultry and other birds that leave the nest early, down feathers are well developed at hatching and are gradually replaced by, and covered with, appropriately coloured contour feathers; the flight feathers develop just before fledging. Nearly all birds have an undercoat of white down beneath their sleek outer feathers and in cold-adapted birds, such as the eider duck that nests during the cold Arctic spring, the down is extensive.

LEFT Palaeontological and embryological evidence agree that tufted down feathers (lower left) evolved first, followed by symmetrical flat-vaned contour feathers (the two feathers upper right). Asymmetrical flight feathers (centre) are the most advanced.

FORMING FEATHERS: AN ORDERLY DEATH

Feathers consist of dead cells filled with a tough, dry, white material called keratin. Feather shape depends upon the arrangement of the cells as they make large quantities of keratin and die. The living cells destined to become feathers form in rudiments in the skin, familiar as 'goose pimples'. The cells divide and mature, or die, under the direction of several messenger molecules produced locally by their own genes. Certain combinations of signals prompt the immature cells to make keratin, harden and stick together before they die, forming rows or sheets, while other signals induce immediate suicide, which forms gaps. Various combinations of these processes produce unbranched strands or branched vanes, whose length and thickness depend upon how many cells participate. If the rudiment starts as a circular blob, its strands of cells form a tuft, as in down feathers; more complicated arrangements align vanes along a central rod, as in contour and flight feathers. The incorporation of special cells that make various coloured pigments as these processes occur adds colour and pattern.

Once they have emerged, feathers can be straightened and oiled by preening but malformation or damage cannot be repaired. Periodic moulting and regrowth replace worn feathers and produce different plumages appropriate to breeding, migration or winter conditions. Producing feathers of appropriate size, shape, mechanical strength and colour in the appropriate places on the body entails very precise regulation of the careers of large numbers of cells. Interactions between several, probably many, control genes ensure that every cell follows a rigid timetable of formation, maturation and death and is marshalled into its place.

Using powerful microscopes and modern methods of detecting when and where genes are active, the biologists have demonstrated that fewer genes were active during the formation of down feathers than are necessary for the development of the more elaborate flat, stiff feathers. The additional genes essential to the formation of contour and flight feathers are not new. They are almost identical to control genes found in all other backboned animals and in many other kinds of animals. In developing birds, their activities are not confined to feather formation: they help build the nervous system, other components of the skin and other important structures. So the inherited change that facilitated the evolution of feathers was not the appearance of new genes, but a shift in when and where genes that produce long-established messenger molecules are active. We now know that many major new structures in complex animals and plants, previously thought to arise from mutations producing 'new' genes, actually arise from redeployment of existing control genes.

BELOW A down feather (left) and small head feather (right) of an African grey parrot, *Psittacus erithacus*, magnified four times. Down is string-like tufts of elongated keratinized cells. Contour and flight feathers are highly ordered sheets of similar cells.

Minor modifications of when and where the control genes are active can generate feathers of many different sizes, shapes, mechanical properties and colours. About a century of selective breeding has produced an impressive diversity of plumage in domesticated pigeons. Similar diverse forms may appear nearly as rapidly in wild populations and, although natural selection eliminates many of them, some survive to breed. Appearance, especially of breeding plumage, promotes speciation by separating breeding populations, even if other structures and habits are very similar. So the flexibility of feather formation has contributed directly to the modern diversity of beautiful birds, currently around 10,000 recognized species.

Genes and fossils reconciled

The involvement of more genes in the formation of stiff, flat flight feathers correlates with the fossils, which show that simple down feathers evolved well before the more sophisticated contour and flight feathers appeared. Their original function was probably thermal insulation, keeping small dinosaurs and the young of larger ones warm. So the earliest feathers had nothing to do with flight. The fossil and developmental evidence suggests that the additional signal molecules needed to build the more complicated vaned feathers did not participate in feather formation in 'feathered dinosaurs' but did so in *Archaeopteryx*.

We still do not know where or when the first flat feathers were formed, or their exact relationship to the evolution of powered flight. The recruitment of additional signal molecules that enabled the formation of more complex feathers may have happened

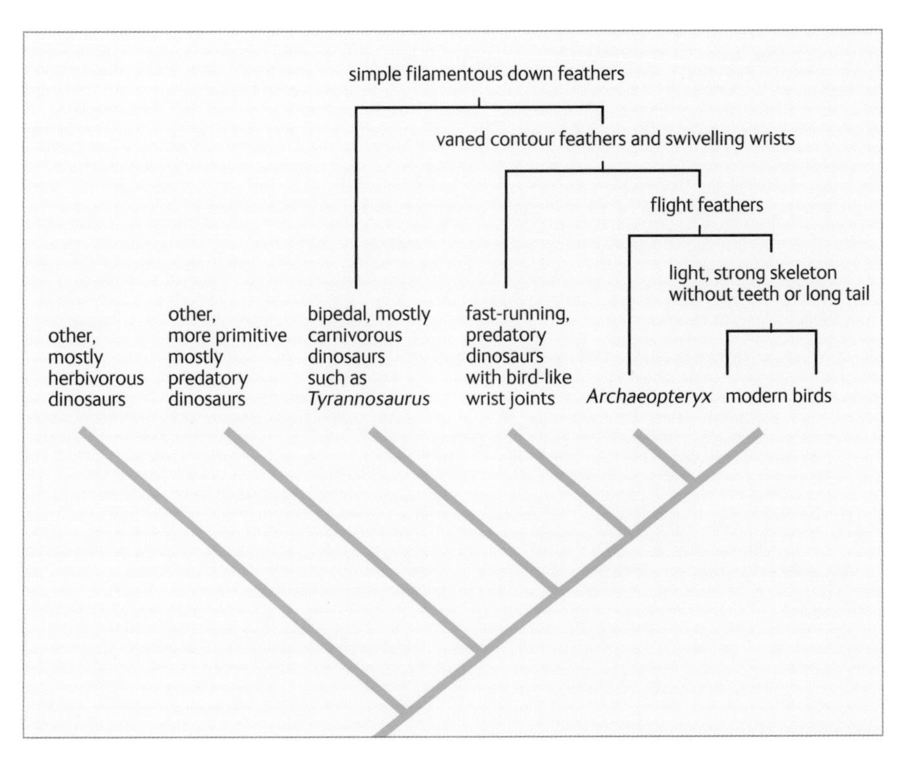

LEFT Tree showing the basic evolutionary relationships between bird-like dinosaurs, *Archaeopteryx* and modern birds. The braces at the top show how various derived characters are shared beween the groups shown.

in other lineages of feathered dinosaurs that are not as well preserved as those of *Archaeopteryx*. None of the Yixian dinosaurs could be direct ancestors of *Archaeopteryx*: it lived 20 million years earlier. In fact, after 150 years as a supreme example of a 'missing link' between major groups of animals, *Archaeopteryx* now looks more likely to be 'uncle' rather than 'grandfather' to modern birds. Its lineage may have disappeared a long time ago, perhaps because it did not acquire the ability to fly by flapping its wings. The absence of a deep keel bone, which in flying birds supports the large, powerful flight muscles, suggests that *Archaeopteryx* probably glided on outstretched wings, or flew weakly, using its long flexible tail as a rudder. Gliding requires very little energy, but is easily disrupted by wind, and is no match for predators capable of powered flight. All known specimens of *Archaeopteryx* (about ten) come from a single rock formation in southern Germany, suggesting that its range was very small, and no obvious descendant species are known. The limited geographic range may explain why.

New structures, same old genes

If down feathers, vaned feathers, flapping flight and all the other features of true birds evolved more than once, defining the boundary between reptiles and birds becomes even more difficult. During the past 30 years, some palaeontologists have abandoned the attempt and classify birds as feathered dinosaurs.

The evolution of feathers illustrates how new structures and new species capable of living in new ways can evolve with surprisingly little change in the genes themselves, which helps to explain how genetically similar species can differ greatly in appearance. It also shows that features that evolved in connection with one function, in this case keeping warm, can be modified to completely different roles, such as becoming airborne. Probably around the same time, shaped and coloured feathers became important in birds' elaborate social and sexual behaviour. The adaptability of feather formation, together with a capacity for long-distance flight, facilitated the diversification of birds, producing the enormous diversity of species on Earth today. Although such non-genetic processes contribute more to diversity than biologists believed 20 years ago, altered or newly formed genes are recognized as the basis for all major changes in structure or physiological conditions. How genes and their interactions contribute to evolutionary change is described in the following chapter, *In the genes*.

OPPOSITE Variations in feather form, size and colour evolve rapidly, producing artificial varieties of domesticated pigeons as well as contributing to the proliferation of natural species.

3
5
4
6
2
1
14
7
13
15
10
11
12
9
8
16
21
19
17
20
34
22
18
24
27
28
29
23
25
26
30
31
32
33
A.F. Lydon

10 CHAPTER *In the genes*

WHEN THE WRITER THOMAS HARDY DIED IN 1928, his ashes were buried in Poets' Corner in Westminster Abbey, but his heart was separately interred in Dorset where it truly belonged. The characters in Hardy's novels are often at the mercy of fate, their stories unfolding in a brooding Dorset landscape that he evocatively captured. In Hardy's world, we are all the playthings of forces that are larger and more enduring than ourselves – victims of circumstance, constrained by history and heredity:

OPPOSITE Thomas Hardy (1840-1928), poet and novelist.

I am the family face;
Flesh perishes, I live on,
Projecting trait and trace
Through time to times anon,
And leaping from place to place
Over oblivion.

The years-heired feature that can
In curve and voice and eye
Despise the human span
Of durance – that is I;
The eternal thing in man,
That heeds no call to die

Heredity, Thomas Hardy

The family face

The "eternal thing in man, that heeds no call to die" is the gene. It is genes that define the family face and trace its characteristic "curve and voice and eye". On occasion, these features are so prominent and distinctive that they are traceable over hundreds of years, like the prominent Habsburg jaw and pendulous lips that were transmitted by intermarriage among 13 families of European nobles and have been traced over 23 generations.

BELOW A stamp portraying King Philip II of Spain, a member of the Habsburg royal dynasty in which a prominent jaw and pendulous lips was a distinctive family trait.

Genes are genetic instructions written in a sequence of molecular letters that are linked together in very long chains of DNA that, with some structural and control molecules, form chromosomes. If we think metaphorically of a gene as being a one-sentence instruction, then a chromosome is like a chapter composed of many sentences. A whole genome is then a book containing many chromosome 'chapters'. In the book of the human genome, there are 23 chromosome chapters. Most cells of your body contains two copies of each chromosome, one copy derived from your mother, the other from your father. If you yourself have children, only one copy of each of your 23 chromosomes is transmitted to them. (If children received both copies from each parent, chromosome numbers would double every generation, from your 23 pairs, to their 46 pairs, to their children's 92 pairs and so on.) The reduction from 23 pairs of chromosomes in your body cells to the single set of 23 that you pass on to your children takes place during the formation of gametes (the collective term for sperm and eggs).

The curious thing about inheritance is that diversity and resemblance both spring from the same source: the genes. Offspring resemble their parents, but they are not identical to them. Diversity among offspring comes from the fact that many genes occur in slightly different versions, called alleles. Alleles arise when mutation changes the letters of the coded messages in the genes. Sometimes such alternations fatally damage the gene's function, and sometimes there is no visible effect; rarely, but most importantly for evolution, mutations may be beneficial for the individual. Though new mutations are relatively rare, the alleles they give rise to are numerous because non-harmful mutations have accumulated over the very long time of evolution. Many, perhaps most, mutant genes are much older than the modern species that carry them and were present in those species' ancestors.

Each of us can have a maximum of only two different alleles for any one gene (one allele on each of the chromosomes in a pair), but in a population as a whole there can be many more than two. For example, the blood groups that determine compatibility between donor and recipient in blood transfusions are determined by three alleles – A, B and O – which all belong to a single gene. Sequencing of the ABO gene using DNA taken from 55 different people discovered more than 80 different mutations hidden within the three alleles, though many of the mutations were neutral, having no obvious effect. Neutral mutations are so abundant in many animal and plant populations that they can be used to create a genetic fingerprint that uniquely identifies an individual from a sample of his or her DNA. DNA fingerprinting is widely used for this purpose in forensic science. When all the genes and all their different alleles are taken into account, we are each a unique combination of our parents' genes; hence each generation delivers family resemblances with a twist of variety.

Chromosome choreography

Sexual reproduction is the key to genetic variation in nature because it shuffles the contents of the genome in three separate ways. Two of the shuffles occur during gamete formation when chromosome numbers are halved in a process called meiosis. First, the two copies of each chromosome inside the cell pair up and may swap lengths of DNA in a process descriptively called crossing-over (or recombination). Afterwards, each chromosome still has a complete set of genes but, where crossing-over has taken place, some of them are alternate versions to those present on that chromosome of the pair before. Recombination is equivalent to swapping sentences between equivalent parts of chapters in two quite different editions of a book.

The second shuffle occurs when paired chromosomes pull apart and separate from each other. One copy of each of the chromosomes migrates towards one end of the cell and the other copy to its opposite end. The original cell divides into two new cells, each now with only a single copy of each chromosome in it. Which chromosome of a pair goes in which direction occurs at random, so the pairing and then separation of chromosomes that occurs during meiosis further shuffles the genome. The process is called independent assortment and is like creating a new edition of a book by random selection of whole chapters from two earlier editions.

The third way in which sexual reproduction contributes to genetic shuffling is when a sperm fertilizes an egg, uniting the chromosomes (23 in humans) of each parent in a new alliance of chromosome pairs. Boy-meets-girl is the part of the sexual process with which everyone is familiar, but not everyone fully appreciates its genetic consequences. A glamorous actress wrote to the intellectual and playwright George Bernard Shaw (1856–1950) proposing that they conceive a child, asking him to imagine how wonderful it would be for the offspring to have "your brains and my beauty". Shaw reputedly replied "but what if he were to have your brains and my beauty?". Bernard Shaw took an interest in genetics and clearly understood that sexual reproduction produces a range of unpredictable outcomes, not all of them adaptive.

BELOW The structure of a genome is like a book composed of many chapters (chromosomes), containing sentences (genes) written with just 4 letters (the letters of the DNA code).

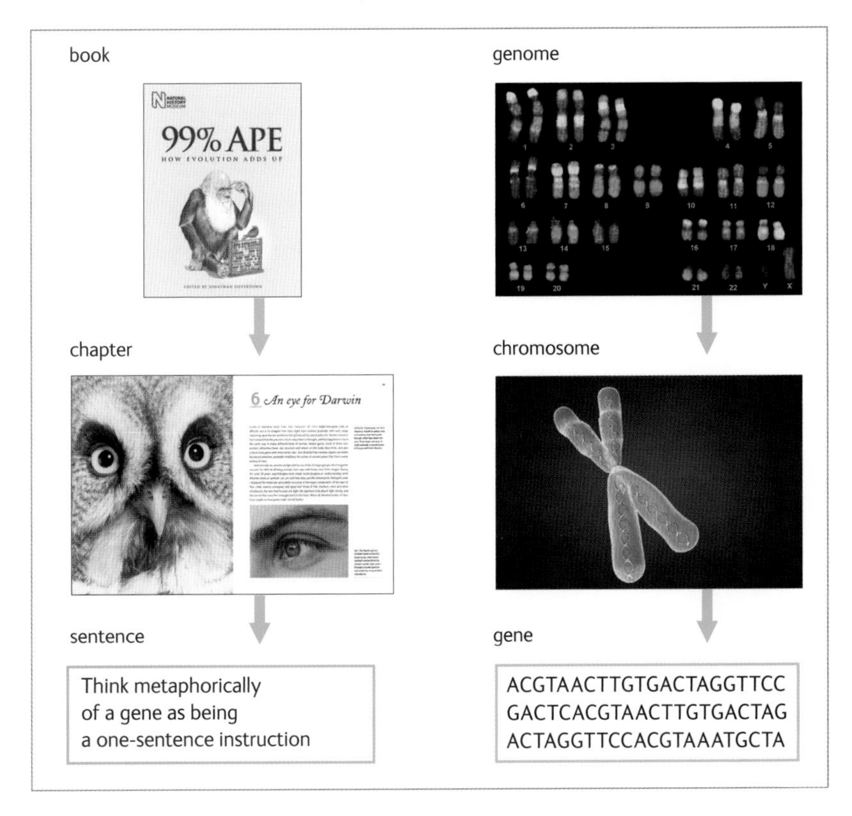

Returning to the metaphor of the genome as a book, let us recap all the ways in which genetics continually re-writes and rearranges the text. The primary source of all genetic variation is mutation. As the book is copied and re-copied, mutations that arise in one copy are passed down to all its descendants. During sexual reproduction, recombination swaps sentences between paired chapters (chromosomes), chapters are separated from one another and through independent assortment are copied into books with new combinations of textual mutation. Finally, whole books are torn in half and chapters are paired again when sperm-text and egg-text unite in a sexual denouement. Is it any wonder that no two books tell the same genetic story?

Genes for every occasion

There are estimated to be about 25,000 genes in the human genome and most characteristics, like height or skin and hair colour, are influenced by some of them as well as being modified by the environment. For many traits, genes do not so much determine the precise fate of the individual as its potential in a particular environment. So, for example, a child with a genetic predisposition to be tall may be stunted in growth by a deficient diet, while fair skin usually darkens with prolonged, repeated exposure to strong sunlight.

The ability of skin to tan is an adaptation that protects against cancer-causing effects of the ultraviolet (UV) radiation component of sunlight, but the response is itself subject to genetic control by a gene called MC1R. About 75 different alleles of MC1R have so far been discovered. Several of them arise from mutations that interfere with the production of the skin pigment melanin and carriers of these alleles have red hair and very fair, often freckled, skin that does not tan. Similar mutations in the MC1R gene have been found in DNA recovered from 50,000-year-old Neanderthal bones in Spain suggesting that, like *Homo sapiens*, some Neanderthals were red-haired and fair-skinned. The Neanderthal genes are different from any of the MC1R alleles found in modern people, so it appears that Neanderthals evolved fair skin independently when the climate in southern Europe was cooler and cloudier than it is now. This repetition is not surprising, since even within our own species, fair skin colour has evolved more than once. Different mutations are responsible for the fair skin of Europeans and east Asians (for example, Japanese).

BELOW Genes controlling skin colour in humans have evolved rapidly, and the differences between populations inhabiting different parts of the globe are quite recent.

In the human population as a whole, genes controlling skin colour show signs of very recent, rapid evolution caused by natural selection. But, if skin pigmentation protects against the harmful effects of UV, why has pigmentation been lost in several populations? There is a rather obvious clue in the geographical variation of skin colour. Skin colour is

darkest in populations living around the equator where exposure to UV radiation is greatest, and lightest in northern latitudes where exposure to sunlight and UV is lowest. Too much exposure to UV is harmful, but too little is also a handicap because UV radiation enables the body to make vitamin D. This vitamin is vital to human health and is particularly important in the development of the embryo; the small amount available in the diet is supplemented by synthesis of vitamin D in the skin, aided by UV light. Thus, natural selection has locally adjusted skin pigmentation, balancing the harmful effects of too much UV against the harmful effects of too little. Where UV is naturally high, this balance leads to dark skin; where UV is low it favours light skin. The balance is very finely adjusted, too. Because of the needs of developing embryos, women require more vitamin D than men, which is probably why women and children in all populations have fairer skin than men.

Vitamin D may be obtained from certain foods and this alternative source can influence skin colour evolution. Inuit in Greenland have darker skins than would normally be expected for a population living so far north, but their traditional diet included seals and fish that are rich sources of vitamin D. Natural selection on skin colour appears to have compensated for this. Modern Inuit who have abandoned their traditional diet and eat supermarket food suffer from record rates of vitamin D deficiency.

Variation in skin colour, eye colour and blood group among individuals are all examples of genetic polymorphism (meaning 'many forms' in classical Greek). We are most conscious (arguably too conscious) of obvious polymorphisms such as skin colour, but they are just the tip of an iceberg of genetic variation. Witness the 75 known alleles of the MC1R gene, nearly all invisible unless you can read a DNA sequence. Polymorphism is the norm in nature and some of the more visible examples in animals and plants provide fascinating insights into the workings of evolution that anyone can observe for themselves. Take a look inside the flowers of practically any species of *Primula* (primroses,

ABOVE Flowers of this primrose have been cut open to show the two forms that occur. 'Pin' (left) has the stigma (female part) at the top of the flower tube and the anthers (male parts) half way down, while 'thrum' (right) has these organs in the opposite locations. The flowers are about three times lifesize.

cowslips, polyanthus), for example, and you can find a polymorphism that Darwin himself discovered: half the plants have flowers with the anthers (male parts) at the top of the flower tube and the stigma (female) near the bottom, while the other half have the male and female organs in the reverse positions. This polymorphism has evolved in several other groups of plants and can be seen in the garden shrub *Forsythia*.

Why be different?

Every polymorphism invites two evolutionary questions. First, what difference (if any) does the polymorphism make to reproductive success? Second, why doesn't one of the morphs replace all of the others? Or, in other words, what balances the polymorphism?

We know the answers for the major genes that influence skin colour, but what about *Primula* flowers? It turns out that the flowers of each of the two morphs only set seed if fertilized by pollen from a flower of the other kind. The matching positions of male and female organs between the different flower morphs increases the likelihood that an insect visiting one type deposits pollen in the correct location to achieve fertilization in the other. This explanation answers the first question and hints at the answer to the second. The reason that neither flower type replaces the other is that each can only reproduce by mating with the other. If one flower type began to increase in frequency at the expense of the other, its chances of reproductive success would fall because of the relative shortage of compatible mates. This process checks any tendency for morph ratios to evolve away from 50:50. The underlying genes that govern the inheritance of flower type also operate in such a way as to always produce an average 50:50 ratio of the two types. Sex ratios in animals (and some plants) are typically balanced at 50:50 in a similar way.

Flower types in *Primula,* and male and female sexes in humans and most other sexually reproducing species, are an unusual kind of polymorphism because every population contains a 50:50 mixture of the two morphs. Other polymorphisms in nature vary geographically, as the genes affecting human skin colour do. For example, the banded snail *Cepaea nemoralis* is highly variable for the colour of its shell, which can be bright yellow, pink or dark brown, and in the number of bands around the shell. Variation in this species, and in its relative *C. hortensis*, has been studied for nearly a century because, long before sequencing genes became possible, banded snails were advertising their alleles upon their shells. The patterns that have been discovered arise from natural selection, as well as other evolutionary processes.

At the continental scale, banded snails with lighter (yellow) shells are most common in southern Europe and darker shell colours become more common in the north, the reverse of the pattern seen in human skin pigmentation. The reason that snail shells show this pattern is that they are affected by environmental temperature. In southern Europe, dark-shelled snails exposed to sunlight for long periods while feeding, for example, may overheat in the warm, sunny climate. In the colder, cloudier climate of northern Europe, having a dark shell is an advantage because it enables a snail to warm up more quickly and to be active for longer. Why, then aren't all snails in the south yellow and all those in the north brown? The reason is that a number of other factors determine colour polymorphisms, illustrating the complexity of evolutionary forces in nature. For example, snails with yellow shells seem to be more resistant to extreme cold, perhaps because, although it sounds paradoxical, some exposed places where it can get very hot in the day can also be very cold at night. Consequently, snails with yellow shells used to be very common in the bottoms of valleys (where cold air collects at night) in rural England. As winters have become milder over the last 40 years, this pattern has gradually disappeared as natural selection has re-balanced the polymorphism away from yellow shells.

ABOVE A sample of animals from a population of the banded snail, *Cepaea nemoralis*, showing the typical polymorphism in shell colour and banding pattern. Snails are shown about half life size.

Another factor at work is predation on snails by birds. Several studies have shown that birds select the more visible shells, which favours patterns that are camouflaged to particular habitats. In woodland where dead leaves and shade provide a brown background,

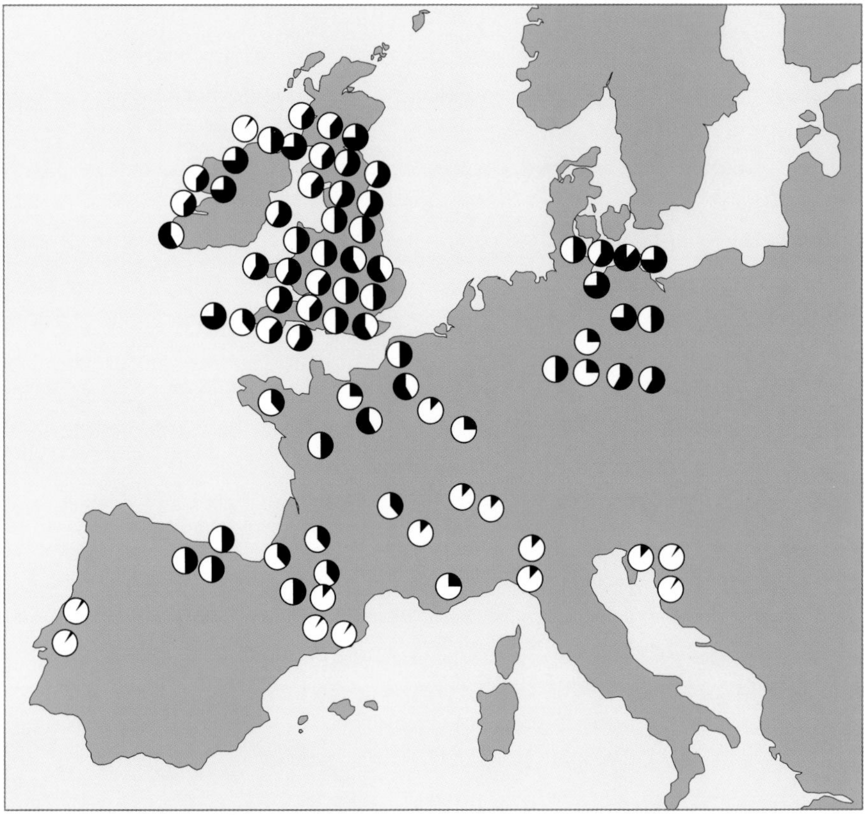

RIGHT Frequency of yellow (white sectors) vs darker (black sectors) shells in populations of the banded snail *Cepaea nemoralis* across west Europe. Darker shells are more frequent in the colder regions of the north.

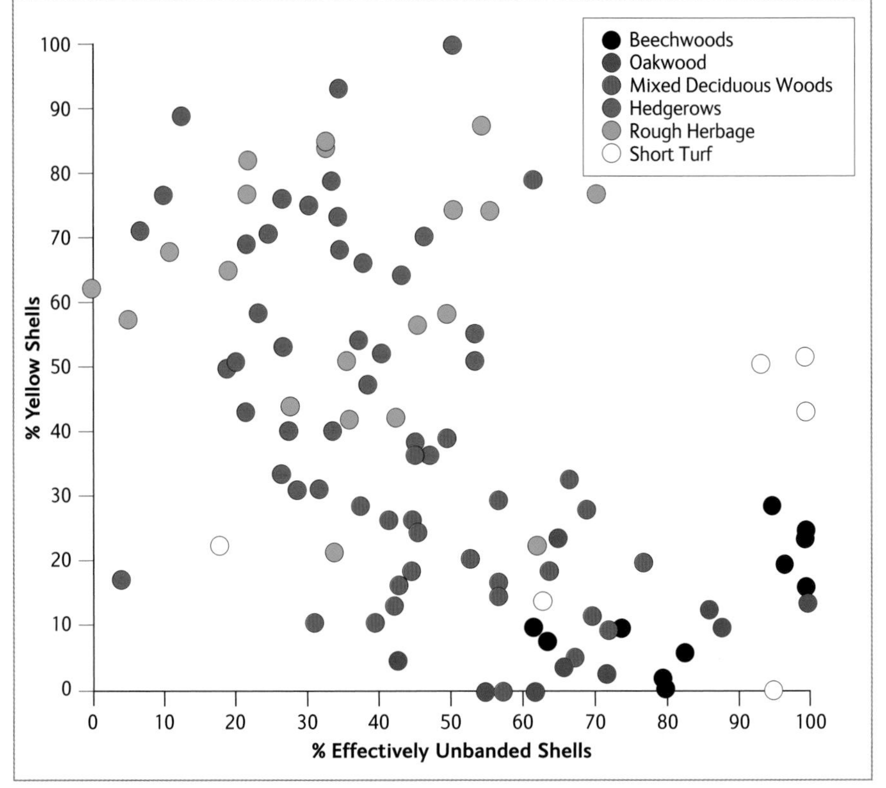

RIGHT Association of shell colour and banding patterns in the banded snail *Cepaea nemoralis* with different habitat types. Yellow, banded snails are most common in open habitats such as grassland and hedgerows, while darker, effectively unbanded shells are most common in woodlands.

LEFT A thrush breaking open shells of the banded snail on an 'anvil' stone. It is possible to study natural selection caused by bird predation on *Cepaea* by comparing which kinds of shells thrushes leave at these stones with the frequency of the different colour and banding morphs present in the *Cepaea* population at large.

darker shells are common. In grassland, yellow, banded snails are better camouflaged and are more frequent. Searching birds tend to miss rarer patterns, which also helps balance the polymorphism.

Finally, snails do move around (albeit slowly!) and are passively carried from place to place. This transport produces patterns of variation that are unrelated to habitat or any of the other ecological factors mentioned. Populations of *Cepaea* in North America, usually founded by the accidental introduction of just a handful of snails brought in with plants from Europe, often consist of only one type of shell pattern. As we shall see in the next chapter, geographical separation of populations can eventually lead to the formation of new species.

11 CHAPTER *New species from old*

LEFT Kate and her foal, on a farm in Colorado, USA. That Kate is indeed the mother of the foal has been confirmed by DNA tests carried out on samples of their hair. The foal's father is thought to have been a donkey.

ALTHOUGH A CHARMING EXAMPLE OF equine motherhood, there's something odd about Kate and her foal: Kate is a mule.

It is popularly supposed that even when closely related species can be hybridized to yield viable offspring, the latter are never fertile, so apparently providing a clear criterion for distinguishing species. Thus mules – artificially induced hybrids between horses (*Equus caballus*) and donkeys (*E. asinus*) – are expected to be infertile. But Kate, rare instance though she is, clearly violates that rule. Hence species are not as clear-cut as they might seem.

On the Origin of... what?

That hybrid infertility is not a hard-and-fast rule should come as no surprise, given Darwin's theory of phylogenetic diversification (Chapter 3 *The tree of life*). According to this theory, species are merely snapshots in time of gradually diverging populations, on the way to forming higher taxa. Yet intermediate stages between fully inter-breeding varieties and discrete species don't seem to be as commonplace in nature as one might expect from Darwin's model of gradual separation. What, then, lies behind the apparent natural identity of species?

The key to this riddle comes from the missing component of Darwin's theory as it stood in his own day – genetics, and especially the controls on how genes are distributed in populations (Chapter 10 *In the genes*). As far as sexually reproducing organisms are concerned, species are populations *within* which genes are freely mixed by interbreeding, but *between* which reproductive barriers of various kinds inhibit the flow of genes. This is known as the 'biological species concept'. Once reproductive barriers have formed, genetic differences between the separated populations can begin to accumulate. Eventually, these differences may become great enough to prevent the two populations interbreeding, even when the original reproductive barrier is no longer present. To understand this process of speciation – the splitting of ancestral species into separate descendent species – we need to look at how these barriers arise. We now understand this process much better than in Darwin's day.

OPPOSITE Male Raggiana bird of paradise (left) courting a relatively drab female (right) with a combination of brightly coloured plumage and ritualized choreography. Elaborate displays such as this are one means by which mating is kept within species.

Brown bear distribution
Polar bear distribution

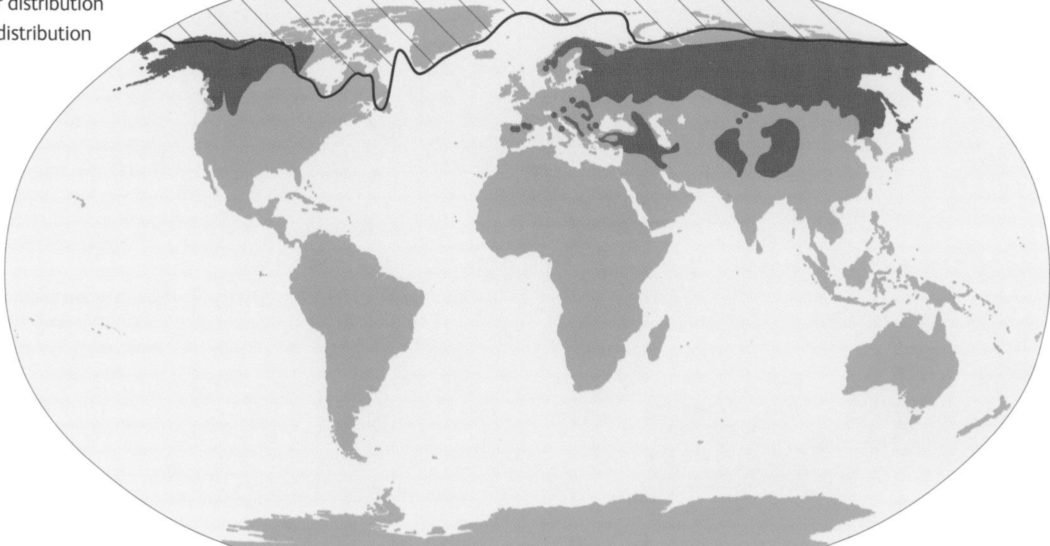

ABOVE Geographical ranges of polar bears and brown bears.

Losing touch

Reproductive barriers emerge from the natural history of the organisms concerned – where and how they live, and especially how they reproduce. Geographical separation of populations is probably the most frequent initiator of speciation. Darwin included this argument in his initial thoughts on the evolution of species (Chapter 3 *The tree of life*), but downplayed it in *On the Origin of Species,* where he placed greater emphasis on the power of natural selection to split populations purely through adaptive divergence between vaguely defined 'varieties' within them. The problem with the latter idea is that it doesn't explain how the all-important reproductive barriers arise. The long-term human experiment of pedigree dog-breeding, for example, shows that even drastic divergence in form and behaviour need not prevent interbreeding when opportunity

BELOW Mounted specimens of polar bear (below), brown bear (far right) and pale brown hybrid (near right) at the Natural History Museum at Tring, UK.

ABOVE Northern parula (left), yellow throated warbler (right) and hybrid between the two species (centre), mistakenly identified in the past as a separate species, called 'Sutton's warbler'.

allows. Without human intervention to control breeding, domestic dogs would soon become mongrelized, so breeds would not separate into different species. How, then, might interbreeding between distinct varieties be prevented in nature?

So long as populations remain geographically separated, interbreeding is obviously denied. Polar bears (*Ursus maritimus*) and brown bears (*U. arctos*), for example, rarely encounter one another in the wild. At least 110,000 years ago (judging from fossil remains), they diverged in appearance and life habits as they adapted to their different habitats: brown bears are omnivorous and sleep for several months in winter, while polar bears are carnivorous, preying on seals throughout the winter. Yet they are still capable of hybridizing, though rarely do so in the wild, and the offspring may be fertile as well as viable (so, strictly speaking, they have not in fact fully speciated according to the biological species concept).

Natural changes in the geographical ranges of closely related species populations can lead to recurrent hybridization. In the eastern United States, a supposed rare species of bird was given the name of Sutton's warbler (*Dendroica potomac*), but was later found to be a hybrid between northern parulas (*Parula americana*) and northward-advancing migratory yellow-throated warblers (*Dendroica dominica*). More such cases can be expected in future arising from the effects of climate change on species distributions.

In time, differences between populations that evolve while they are separated may lead to genetic incompatibilities between them. There are many ways in which such incompatibilities might arise, such as a change in the numbers of chromosomes or the arrangement of genes on them. As a consequence, the parental sets of chromosomes inherited by a hybrid don't match up (for example, horses have a total of 64 chromosomes, and donkeys, 62). Although a hybrid may be viable, the mismatch of its chromosomes produces defective gametes at meiosis. Thus when populations reconnect, an effective barrier to gene flow between them already exists.

SPEEDY SPECIATION

Cichlids are a widespread and spectacularly diverse group of tropical freshwater fish. Some 600 species live in Lake Malawi, in East Africa, almost all of them endemic (that is, they occur nowhere else), including at least 295 species that typically inhabit the rocky shores of the lake, feeding in a wide variety of ways, and collectively referred to by local people as '*mbuna*'. Both genetic and geological evidence points to extraordinarily rapid (for a vertebrate animal) speciation among them.

Sedimentary deposits show that the size and water level of Lake Malawi has fluctuated markedly throughout its 4–5 million year history. From 1.6 million years ago until between 1 and 0.57 million years ago, the lake virtually dried out. Hence the evolution of the cichlid species flock must post-date this period. Changing climatic conditions later caused significant fluctuations in lake levels, instigating waves of extinction among its inhabitants followed by re-diversification. The latest phase of speciation is thought to date from as little as 11,000 years ago, when a warmer, wetter climate filled the lake again, creating new shallow water habitats separated by deeper-water areas. Many *mbuna* species, though obviously different in colour, size, diet and breeding habits, turn out to have little difference in DNA sequences, which is consistent with such recent speciation. Behavioural studies show that different forms breed only with each other, and so are true species, not just varieties like artificial breeds of dog.

The observation that certain species are found only in the waters around certain islands that formed around 1860, having previously been dry land, has suggested that some may have arisen only since that date, though this interpretation is controversial.

ABOVE Some examples of 'mbuna' cichlids from Lake Malawi.

Moreover, there may be natural selection against poorly adapted or less viable hybrids. Such changes take time to evolve, which is why speciation happening this way cannot be observed directly and has to be inferred from the remaining evidence – the presence of similar DNA sequences in the different populations, fossils or other geological evidence for the changing circumstances of ancestral populations. In a few instances, such evidence suggests that speciation occurred in only thousands or even hundreds of years. Geographical isolation can come about in many ways. Rare chance dispersal may found isolated populations, as with Darwin's finches, another example of rapid evolution (Chapter 12 *Darwin's finches*). Over centuries to thousands of years or more, climatic changes may create environmental barriers between formerly interbreeding populations, such as deserts, ice-sheets, rivers or deep-water areas in lakes. Over longer time intervals, continental drift may create major barriers as continents become separated by new oceans or collide to form mountain chains.

Knowing your own

Producing gametes, especially eggs, can be expensive in terms of energy and resources, so natural selection tends to eliminate their wastage on any mating that is unlikely to promote reproductive success. Where fertilization is internal and animals have well-developed sense organs and complex behaviour, as in vertebrates and insects, adaptations that ensure recognition of suitable mates of the same species often involve distinctive appearance and elaborate courtships. Sexual selection, usually by females of males, can further hone these ritualized performances to an extraordinary degree.

While such behaviour serves to keep mating within species, divergence in mating rituals acquired while populations are separated may inhibit hybridization when they again live in the same areas, precipitating a pre-mating barrier between them. Some groups of organisms are highly susceptible to speciation in this way when their populations have been separated even for relatively short periods. In the Hawaiian chain of islands, there are some 800 species of fruit flies (drosophilids), over 95% of which are endemic. Detailed studies of the genes of these endemic species suggest that they have all descended from one species, perhaps even a single fertilized female that arrived on one of the islands from North America a few million years ago. Many closely related species of these flies are genetically and ecologically similar, but differ in characters connected with their elaborate courtship dances, such as head shape and wing patterning. It thus seems likely that sexual selection has played a major role in this proliferation of fly species, as well as in the evolution of the cichlid fishes in the Great Lakes of East Africa (see box).

BELOW A picture wing fly on a leaf in Hawaii, showing the distinctively broad head and wing pattern that are important for courtship in this species, *Drosophila heteroneura*.

No sex, please, we're budding

Many plants and animals can reproduce without sex, but among bacteria asexual reproduction, involving simple replication of their single loop of DNA and cell fission, is the norm. The result of this process is clones, or large families of genetically identical individuals. Clones differ from one another only through rare mutations, each of which establishes a new clone. Despite the rarity of such mutations per individual bacterium, the sheer numbers and rate of reproduction of bacteria nevertheless ensure a continuous and abundant supply of new clones.

Many bacteria also exchange genes with each other, though in a less precisely choreographed way than in true sexual reproduction. Exchanges usually occur between closely related lines, though not invariably so. Combined with their prodigious rates of reproduction, such exchanges help disease-causing bacteria, for example, to evolve resistance to antibiotics with worrying rapidity (Chapter 15 *Catch me if you can*). Here, 'speciation' is really a matter of gradual separation of similar clonal races through the extinction of less successful intermediates between them. However, given the lack of regular sexual reproduction and somewhat haphazard exchange of genetic material between individuals, we cannot really compare such bacterial 'species' with those defined in terms of reproductive barriers as above.

Fossils and species

What fossils can tell us about speciation is limited for two reasons. The first and most obvious reason is the scant evidence they can furnish for fertility, reproductive barriers, gene flow and the like. In exceptional cases – limited to very recent fossils – DNA fragments can be recovered in good enough condition to be analysed, which might then throw some light on these aspects. This limitation means that it is almost impossible to apply the biological species concept directly to fossils, though species can be inferred by comparing the variability recorded in fossil samples with that seen in extant species.

The second problem is the incomplete nature of the fossil record (Chapter 3 *The tree of life*), not only through time, but also in geographical extent, both because of patchiness in where sediments accumulated and their limited exposure now. Nevertheless, the search for unusually complete sedimentary successions over the last half-century or so has yielded many detailed evolutionary case histories. Favoured sites include lake deposits that accumulated steadily in subsiding rift valleys and sediments that built up over long periods in deep marine basins.

Long-term change in the characteristics of species, including reversals, can be illustrated among trilobites, an extinct group of mostly sea-floor dwelling animals that superficially resemble woodlice. Sampling from strata that were originally deposited in a deep marine basin in what is now central Wales has provided a detailed picture of gradual transitions between species in eight distinct evolutionary lineages. The longitudinally three-lobed body (from which trilobites acquire their name) shows a clear division into a

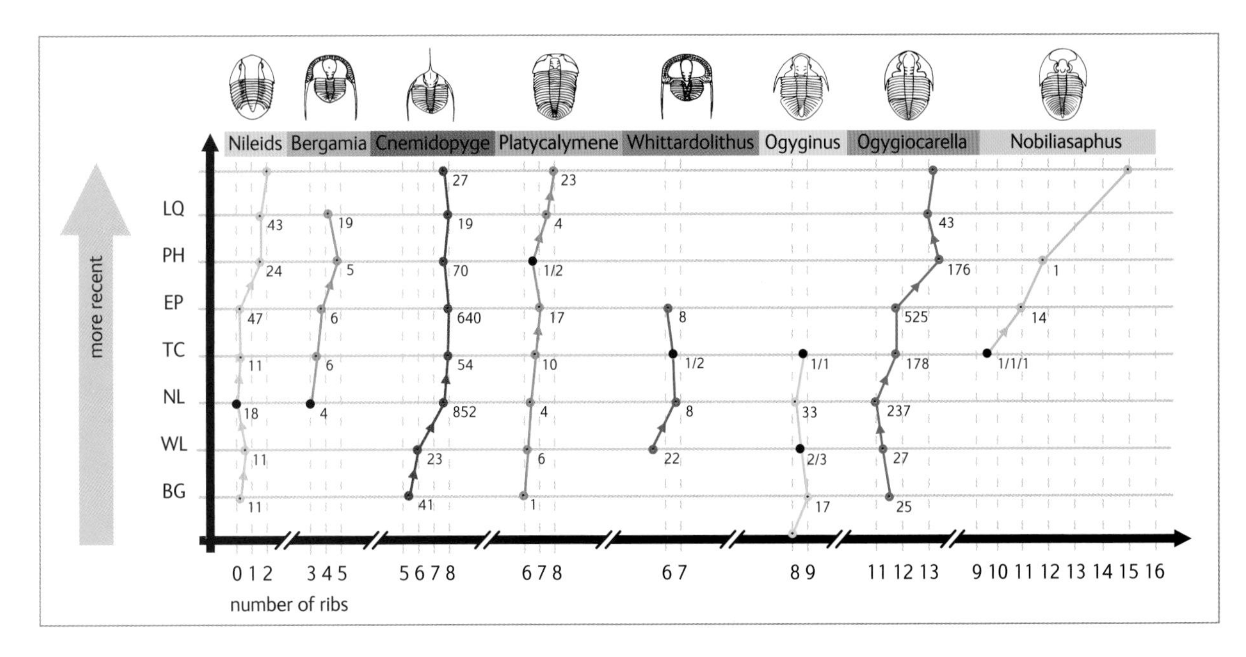

ABOVE Changes in rib numbers on the tail-pieces of trilobites in successive samples from eight lineages in Ordovician strata in central Wales. Numbers next to the data points are sample sizes, horizontal lines through them show 95% confidence intervals and small arrow-heads indicate statistically significant changes. The letters at left (BG, WL, etc.) denote sections of sampled strata. The chronological sequence is upwards and the seven sections span in total about 2 million years.

shield-like head, an originally flexible segmented middle part and a tail-piece consisting of fused segments. The number of ribs on the tail-piece is one of the characters that reveal these gradual evolutionary changes. However, there are in addition numerous well-documented examples of species that show little or no evidence for longer-term change, a condition known as stasis. Stasis can nevertheless be broken by occasional, rapid 'punctuational' shifts.

Such fossil patterns allow us to draw three general conclusions:

- Species are capable of evolutionary change through time (thus refuting the only testable part of the hypothesis of special creation, as noted in Chapter 3 *The tree of life*).
- Such change *can* be geologically rapid – on the scale of centuries to millennia – which is consistent with inferences based on some living species, as discussed above.
- Nevertheless, many species show prolonged periods of stasis.

The frequency of stasis is perhaps the most interesting discovery, as it would not necessarily have been predicted by Darwinian evolutionary theory (though it is not inconsistent with it). There are several possible explanations. The most likely one is that species that are well adapted to a given set of environmental and ecological conditions are preserved by natural selection. They do not change because mutations are all deleterious. Whatever the reasons for stasis, it may help to explain why most species appear to us to be more or less distinct and stable natural entities, even if various degrees of hybridization between related species sometimes make the boundaries between them seem a little fuzzy.

One of the events that commonly triggers speciation is migration into new environments. Thus, rapid speciation is often seen in populations that have recently colonized islands. The most well-known and well-studied example of this is in Darwin's finches in the Galápagos Islands.

'Unfortunately most of the specimens of the finch tribe were mingled together; but I have strong reasons to suspect that some of the species of the sub-group Geospiza are confined to separate islands. If the different islands have their representatives of Geospiza, it may help to explain the singularly large number of the species of this sub-group in this one small archipelago, and as a probable consequence of their numbers, the perfectly graduated series in the size of their beaks.'

OPPOSITE A mocking bird, *Nesomimus macdonaldi*, on the island of Española. There are three other endemic species of mocking bird on Galápagos. Darwin called them 'mocking-thrushes'.

By the time *The Voyage of the Beagle* was published (in 1845), the finches had assumed a much greater importance, but at the time he collected them he did not even appreciate that they were all finches and the entire collection from two islands was placed in one bag. This should not be seen as criticism, for at the time he had no reason to suppose that the diversity of these birds was any different from the diversity on the nearest coast of mainland South America, which the *Beagle* had not visited. In any case, he was going to work on the collections back in Britain. Although he did not appreciate their significance at the time, he and the other collectors still made a comprehensive collection of the birds and the specimens excited interest as soon as the collection reached the Zoological Society of London on 4 January 1837. Within a few weeks, John Gould was describing to the Zoological Society his first observations on a new group of species of ground finch.

ABOVE The large ground finch, *Geospiza magnirostris*, photographed on Santiago island in the Galápagos archipelago.

LEFT A plate drawn by John Gould for the Birds section of *Zoology of the Voyage of H.M.S Beagle*, edited by Charles Darwin and issued in 19 numbers. The plate shows the male and female large ground finch, *Geospiza magnirostris*. The female has a lighter plumage.

Beaks and seeds

BELOW The species of Darwin's finches showing how the beak type is related to the food they feed on. The finches are grouped according to the action of the beak, for example, crushing or probing.

The beaks of Darwin's finches show a surprising variety of shapes and there is a strong correlation between the size of the beak and the size of seed that the beak is able to crack. However, it is not just the differences between the species that are of interest. Within a species there can be gradation of beak size. For example, in a population of the medium ground finch (*Geospiza fortis*) studied by Peter Grant on Daphne Major in 1976, the depth of the beak varied from 7.3 mm to 10.8 mm (0.29 to 0.43 in). There was a severe drought the following year, seeds were in short supply and as the birds

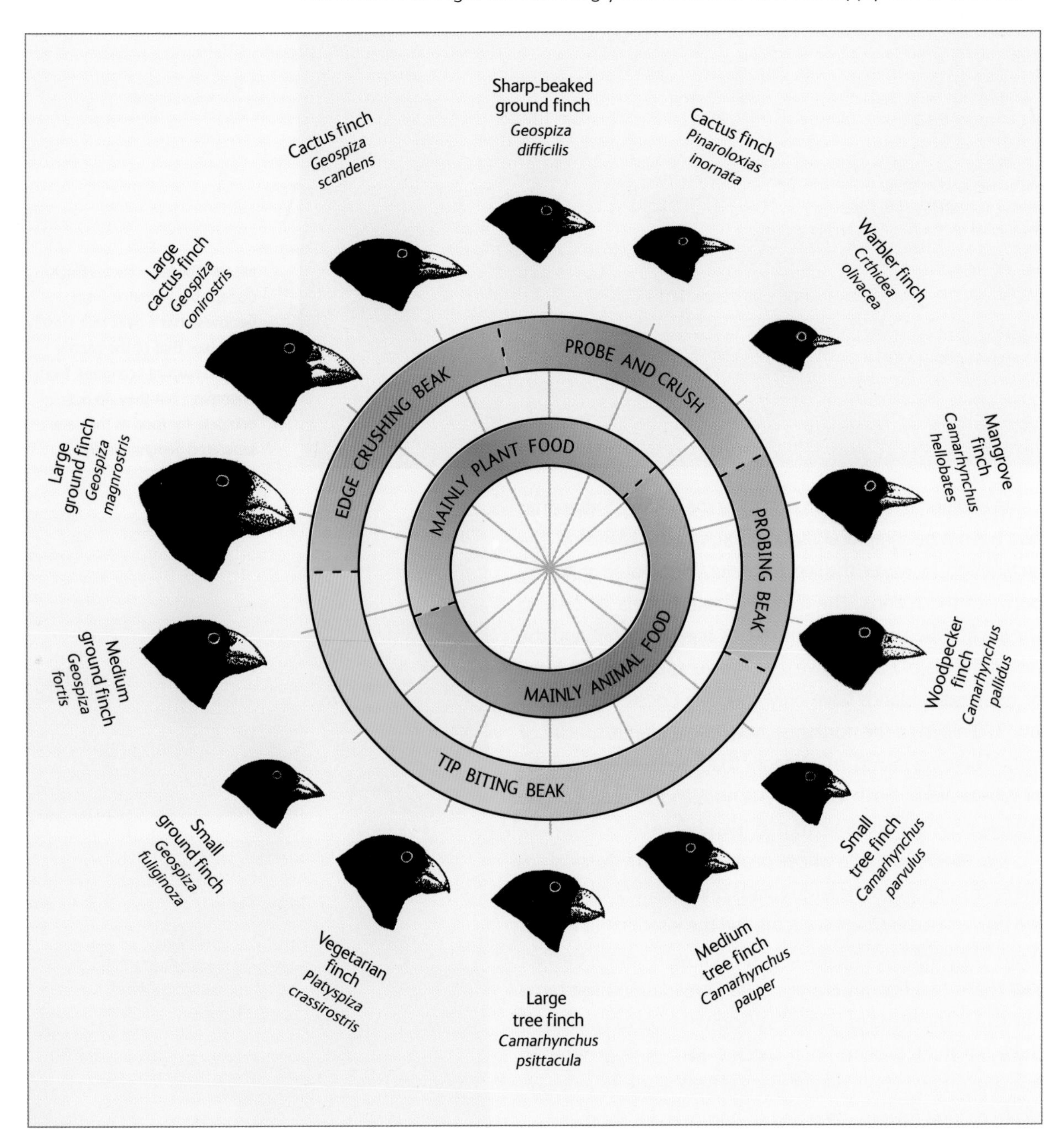

foraged on the ground, seeds that were easy to eat became progressively rarer. There were still seeds of the caltrop (*Tribulus*) but these were hard to crack. The larger-beaked birds were at a small advantage.

Proportionately more of the larger-beaked birds survived the drought year and the mean beak size of these survivors was nearly 1mm (0.04 in) larger than the mean had been for the whole population. The sex ratio was also unusual in this group of survivors in that there were far more males than females. This might be predicted as the males are generally larger than the females, have larger beaks and so can crack larger seeds. So, the weather conditions produced natural selection in favour of larger beaks. It was when the offspring of birds that bred in 1976 were compared with the offspring of birds who survived the 1977 drought to breed in 1978 that evolution could be seen in action. The mean beak depth of the 1978 offspring was larger than the 1976 offspring. Thus a change in beak depth of the population brought about by selection in 1977 had been translated into a larger mean beak depth in the next generation. There had been an inherited increase in beak depth.

ABOVE Seeds of the caltrop, *Tribulus* cistoides have hard woody tissue that protects them. Large finches can crack the case open to get at the seeds, but smaller finches find it difficult or impossible.

Beaks and genes

The genes that underlie the variation in beak depth are largely unknown but research has recently identified two significant indicators of genetic control of beak size. There are a number of growth promoters that appear around the beak while the embryo is developing within its egg, including a molecule called bone-promoting protein 4, or Bmp4 for short. In the large ground finch, this protein can be found over a larger area of the developing beak and at higher concentrations than in embryos of species that are related, but have smaller beaks. The protein also appears earlier in development in the large ground finch. In general, it appears that species with deeper, broader beaks, relative to their length, show a greater amount of Bmp4. The amount of Bmp4 present is under genetic control, as can be demonstrated by experiments in which the amount of Bmp4 produced is artificially increased by changing the expression of the genes. When this experiment was carried out in chick embryos, it resulted in chicks with broader and deeper beaks.

A second indicator of genetic control is another molecule, calmodulin (CaM), which has a complex role in the control of calcium. Calcium, in turn, is involved in a range of processes in cells and tissues. The production of CaM is higher in cactus finches, with their long pointed beaks, than it is in ground finches, with their shorter more robust beaks. The suggestion is that a change in this and other genes producing proteins that regulate beak development is the primary molecular change brought about by selection.

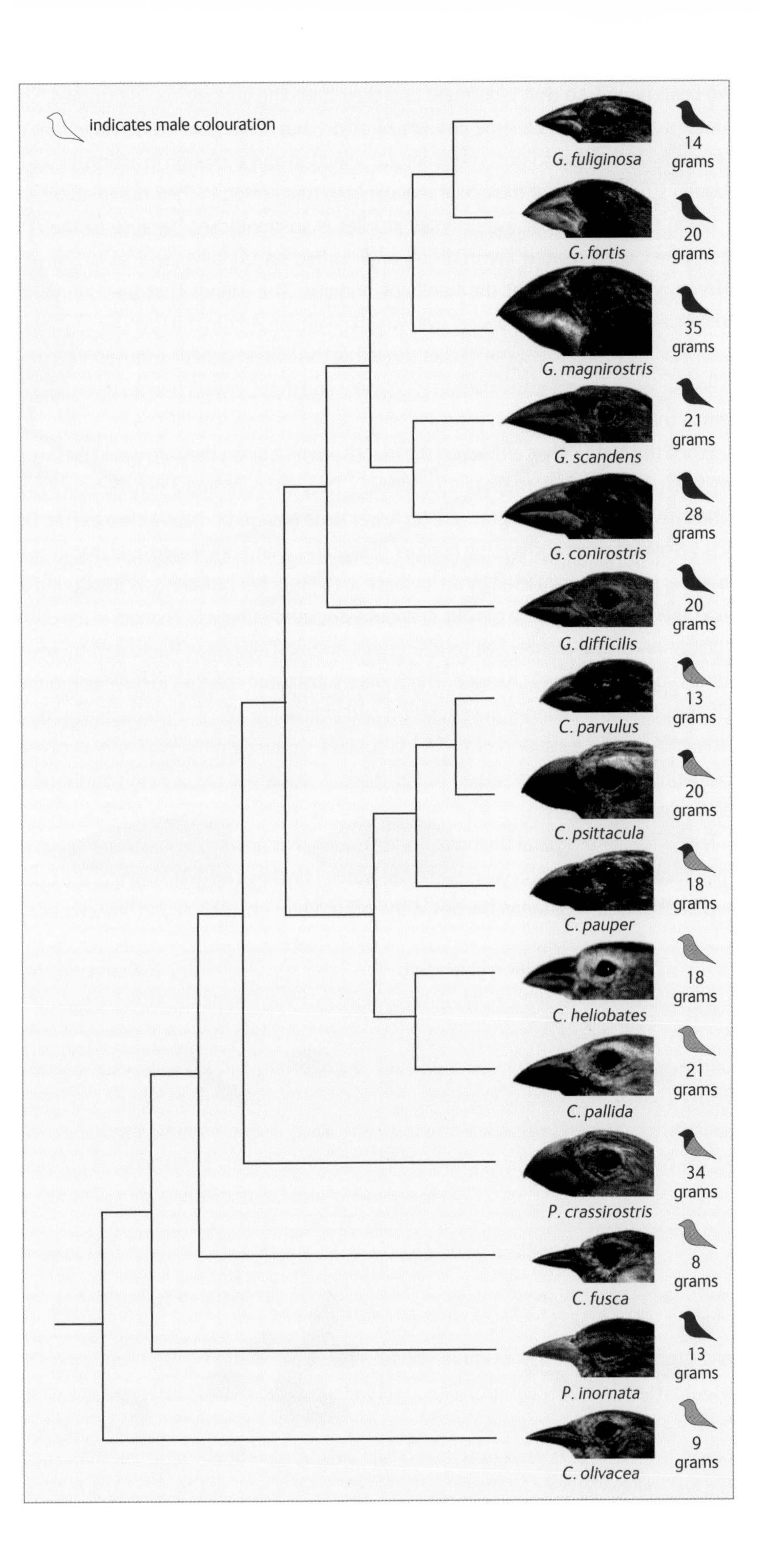

RIGHT DNA markers can be used to determine genetic relationships in the species of Darwin's finches. The finches are all very similar genetically and so the relationships shown in this DNA-based ancestral tree are best estimates. Nevertheless, there is fairly good agreement between the groupings of species, and the groupings are determined by traditional taxonomic study.

more 'bushy', with the appearance of ground and tree finch species. Each speciation event would have started with the separation of two populations of the same species, probably as a result of a small group of individuals colonizing another island. Colonization is a rare event, given that finches are weak fliers. The accumulation of changes in each population brought about by natural selection would, over time, lead to the formation of two species. The islands of Galápagos are quite variable, offering different types of environment, and supporting different combinations of animals and plants. Also, volcanic eruptions and changes in sea level are forming more islands. Both these factors have provided favourable conditions for increased speciation.

The genetic analysis that has made it possible to produce the new phylogeny of the finches will be improved to refine understanding of the relationships between species and perhaps to show individual populations within species. We might eventually be able to see the stages leading to future speciation. The Galápagos finches have become the premier example of evolution in action and it is fitting that this example should have come from the collecting bag of Charles Darwin himself.

When Charles Darwin had returned from the Galápagos Islands and was settled at Down House in Kent, he wrote to his friend Joseph Hooker, Director of Kew Gardens, seeking his advice and modestly denying any knowledge of botany. However, in his later years Darwin studied plants extensively and wrote books about his researches on orchids, insectivorous plants, climbing plants, flowers and plant reproduction. In his autobiography, he said:

> ‘*It has always pleased me to exalt plants in the scale of organized beings.*’

It is therefore fitting that whereas Darwin's finches illustrate the minutiae of natural selection and evolution, the flowering plants live up to Darwin's exalted expectations of them and show evolution at work on a very grand scale.

RIGHT *Mimulus cardinalis*, crimson monkeyflower, by Arthur Harry Church, 1903.

13 CHAPTER *The flowers of evolution*

LIKE ANIMALS, THE PLANT KINGDOM ALSO HAS marine ancestors, but unlike animals, plants only made one successful invasion of the land. Hence all land plants, from mosses to giant redwoods, are descended from a single terrestrial species of pioneer. There is little doubt that the marine ancestors of the land plants must have been green algae, because (among other things) these algae share with land plants the same vital, miniature passenger tucked inside their cells that gives them all the fundamental characteristic of plants: the ability to synthesize their tissues from the simplest ingredients and the most widely available energy source. The almost magical chemistry of photosynthesis, that uses the power of sunlight to synthesize sugars from carbon dioxide and water, takes place inside tiny green bodies called chloroplasts. As well as fuelling living processes, these sugars assemble in various ways to produce cellulose, the main structural material of terrestrial plants.

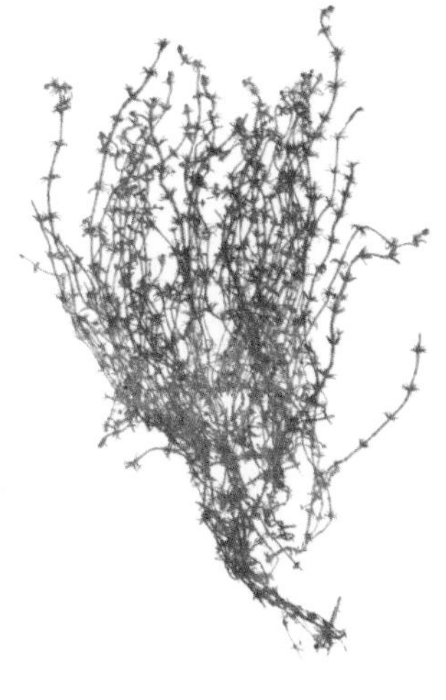

ABOVE *Chara desmacantha*, lesser bearded stonewort. Stoneworts are believed to be the closest living algal relatives of land plants.

First plants

Curiously, chloroplasts have their own DNA, which occurs in a circular chromosome similar to that found in bacteria (Chapter 3 *The tree of life*) and mitochondria. In fact, it appears that chloroplasts *are* remnant blue–green bacteria (cyanobacteria) that were once free-living but are now confined within plant cells that captured them long ago from the sea. The origin of chloroplasts from free-living cyanobacteria is an example of symbiosis, from the Greek for 'living together'. Symbiosis is an important catapult of evolutionary change, producing new combinations of biological function that can open up wholly new pathways of evolution. The new use for existing parts that occurs in symbiosis is another instance of how evolution creates novelty by combining pre-existing components (Chapter 6 *An eye for Darwin*). Many of the genes vital to photosynthesis that were acquired through symbiosis have been transferred to the plant's nuclear DNA, leaving the chloroplast with only about 5% of the genes found in free-living cyanobacteria. Thus robbed of most of their genes, chloroplasts are entirely dependent upon their hosts and are only able to divide and proliferate inside host cells.

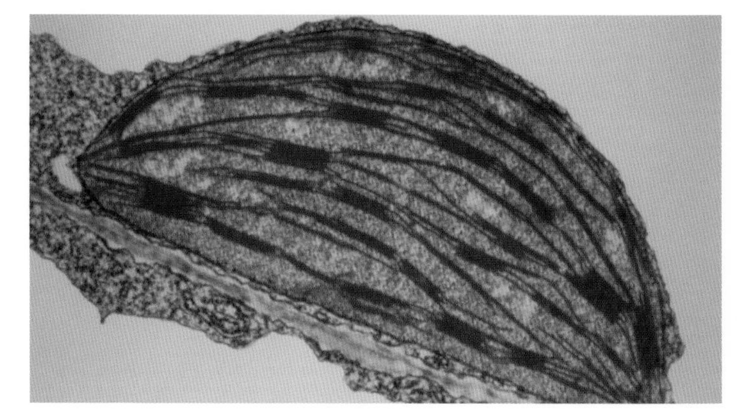

BELOW A false-colour transmission electron micrograph of a chloroplast from a tobacco leaf, *Nicotiana tabacum*.

RUSSIAN DOLLS AND MALARIA

Quite separately from green algal evolution, another ancient photosynthetic cyanobacterium formed its own symbiosis with the eukaryotic ancestors of red algae. Like a collection of Russian dolls, some of these red algae were in their turn engulfed and became symbionts of other single-celled organisms that evolved very un-plant-like careers. There is the lingering fragment of a red algal symbiont in the malarial parasite, *Plasmodium*. This blood parasite, which kills more than a million people annually, has dispensed with the genes for photosynthesis altogether, but kept remnants of the chloroplast containing other useful genes. Without these genes of chloroplast origin *Plasmodium* cannot thrive inside host cells, but the hosts themselves are not dependent upon such genes, suggesting that they might be a target for anti-malarial drugs that would kill the parasite without harming the host.

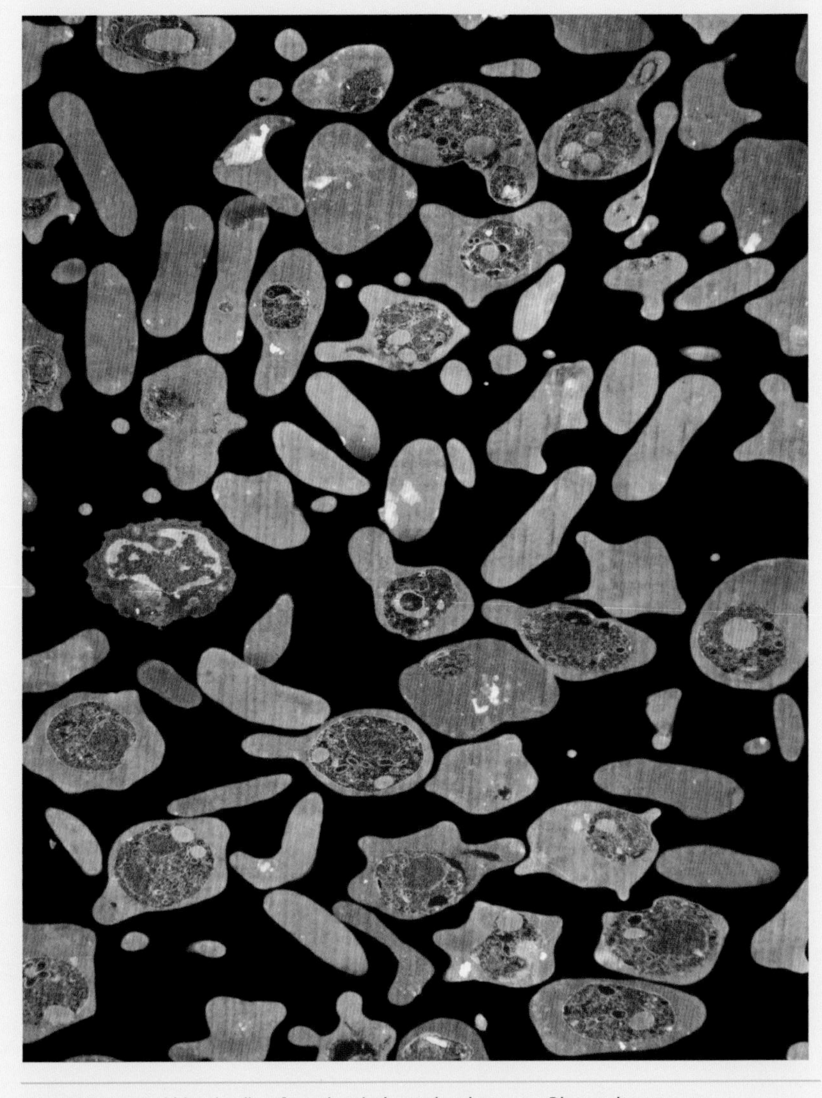

ABOVE Human red blood cells infected with the malarial parasite *Plasmodium* sp.

Evolution, being a slow and gradual process, almost never comes up with something completely and utterly new, but continually surprises us with ingenious new uses for old devices, like the chloroplast genes that have turned up in the malaria parasite (see box). New symbioses are rare in evolutionary history, but, as in the origin of the green algae and the land plants that evolved from them, symbiosis can enable life to colonize new habitats. Animal life as we know it would be impossible without land plants, the major food source for them or their prey.

BELOW A liverwort belonging to the group that is sister to all other land plants. Approximately life size.

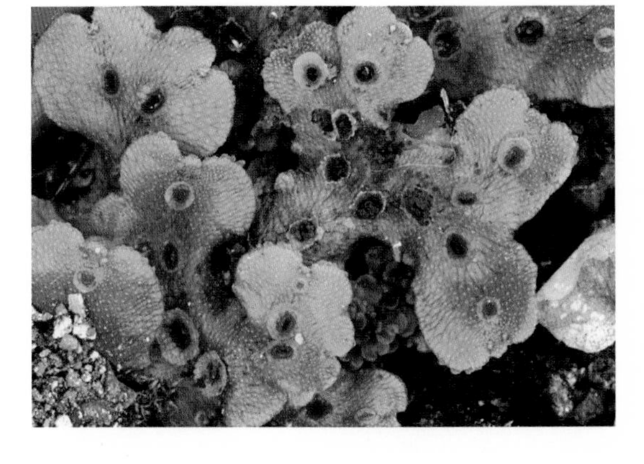

The greening of the land

Just as we humans are naked apes, land plants are really terrestrial green algae. Evolving new tricks, however remarkable, does not sever a group from its origins. Evidence of fossil spores and estimates from the molecular clock (Chapter 5 *African genesis*) suggest that plants first appeared on land about 490 million years ago in the Ordovician. A molecular phylogeny of the land plants shows liverworts, which reproduce by spores, to be a sister group to all other land plants. This conclusion does not mean, of course, that the earliest land plants *were* liverworts, but it does seem that, like liverworts, the earliest fossil land plants were small, spore-producing and simple in structure. *Cooksonia*, for example, which occurs in many fossil localities in North America and Europe, carried its spore-producing structures on simple, branching stems that were only about 6.5 cm (2.6 in) tall. The earliest examples of *Cooksonia* are found in Middle Silurian rocks (425 million years ago) in Ireland.

Life on land presents obvious challenges to any sea creature, chief among them being how to protect eggs, sperm and embryos from drying out. Land plants share one innovation that enables them to reproduce away from water: unlike their marine ancestors, all land plants protect their eggs and resulting embryos within the tissues of the mother plant. But not all land plants have freed themselves from other aspects of dependency upon water. Though they protect their embryos successfully, ferns, mosses and liverworts all require an environment that is wet enough at some point during the life-cycle to enable sperm to swim through a film of moisture to the eggs. These plants are like amphibians in the animal kingdom, able to survive on land but dependent upon wet conditions for reproduction. However, seed plants, which include cycads, ginkgos, conifers and flowering plants, have overcome this limitation and are fully land-adapted.

The earliest fossil seeds date from near the end of the Devonian, 365 million years ago. The seed is

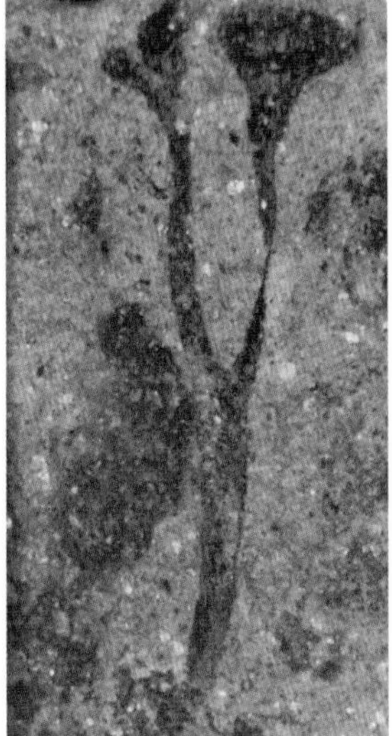

ABOVE AND LEFT Fossil *Cooksonia*, around 10 mm (1/3 in) in length, and a reconstruction of this tiny, early land plant.

OPPOSITE A fly agaric toadstool, *Amanita muscaria*, that forms a mycorrhizal association with tree roots.

definitely a terrestrial innovation, although its derivation from aquatic ancestors can be seen in how the eggs of the most primitive extant land plants, cycads and ginkgos, are fertilized. These plants have mobile sperm that swim the last part of their journey between the pollen grain and egg through a fluid medium by synchronized lashing of numerous hair-like flagellae (singular: flagellum).

Marine algae either float free or, like the familiar seaweeds of the seashore, attach themselves to rocks by a structure that is appropriately named a 'holdfast'. Only land plants have true roots that take up water and nutrients from the soil as well as providing anchorage. The roots of nearly all land plants have a symbiotic relationship with fungi that is crucial to their nutrition. Most of the mushrooms and toadstools you see growing on the ground in woods and fields are part of fungi that are symbiotic with trees or other plants and betray the presence below ground of a dense, pervasive web of fungal threads. These so-called mycorrhizal fungi help supply mineral nutrients and water to their host plants and may also help protect them from infection by harmful bacteria and other pathogens. Mycorrhizal associations between land plants and fungi appear to be ancient. Liverworts, sisters to all other land plants, are mycorrhizal. The association with fungi may therefore have been instrumental in helping the green algae colonize land.

The most diverse of land plants are the flowering plants (angiosperms), of which at least 250,000 species are estimated to be living today (the actual number may be nearly twice as many). Flowering plants evolved from other seed plants called gymnosperms.

LEFT *Ginkgo biloba*, an ancient gymnosperm.

RIGHT Pine forest in Sierra Nevada, California.

ABOVE A cycad, an ancient gymnosperm.

This group of about 700 living species includes the conifers, the cycads, the ginkgos, and some others, but it seems likely that the closest gymnosperm relative of the angiosperms is extinct. If so, the lack of information about its genes would explain why it has proved difficult to date the split between the gymnosperms and angiosperms using a molecular clock. Most such estimates date this divergence to between 346 and 367 million years ago (Late Devonian to Early Carboniferous), although the true date may be more recent than this.

The rise of the flowering plants

Writing to the botanist, Joseph Hooker, Charles Darwin called the origin of the angiosperms "an abominable mystery". The earliest evolution of the angiosperms is still a matter of conjecture, but giant strides have been made in the last fifteen years in understanding how the flowering plants evolved after they had split from gymnosperm ancestors. Indeed, thanks to spectacular success in the use of DNA sequences in tracing how the green branches of the tree of life are connected, the angiosperms are the first major group of organisms to be re-classified using this evolutionary evidence (see box opposite). Perhaps for the first time since Darwin's own day, botany is in the vanguard and not the guards' van of science, and it is appropriate that it should be so, since the picture revealed is one in which the evolution of flowering plants set the pace of evolution for many animals.

NEW PATTERNS AMONG THE GREEN BRANCHES OF THE TREE OF LIFE

Certain natural plant groups have been recognized by botanists and herbalists for centuries. The daisy family (Asteraceae), the carrot family (Apiaceae), the pea family (Fabaceae) and the grasses (Poaceae), for example, all form obvious groups. DNA analysis confirms that these families are indeed distinct groups in the tree of life, separated from others on the ends of long branches in the phylogeny of flowering plants. But there have been surprises from DNA analysis too. For example, the group of colourful, evergreen New Zealand shrubs called hebes, which are popular garden plants, turn out to have sprung from within the speedwell genus *Veronica*, which are mostly herbaceous and not woody. So, technically, hebes ought to be re-named veronicas. Evolution on island archipelagos like New Zealand quite often produces novelties that can turn out to have unlikely seeming affiliations with mainland ancestors. This is because big changes can evolve rapidly when colonists arrive in a new area that has few or no competing species (Chapter 12 *Darwin's finches*). The same can happen on continents when environmental change produces new challenges and opportunities.

Another surprise uncovered by DNA analysis is the closer-than-realized relationship between tomatoes and potatoes, which used to be placed in separate genera within the nightshade family (Solanaceae). Tomatoes, aubergines and potatoes are now all recognized as members of one large genus called *Solanum* that contains about 1500 species. The most radical discovery of all emerging from the DNA-based phylogeny of flowering plants is that the distinction that has long been made between dicotyledons (plants whose seedlings start with two 'seed' leaves) and monocotyledons (plants whose seedlings emerge with only one seed leaf) is oversimplified.

Dicotyledons ('dicots' for short) are a ragbag of broad-leaved plants containing several groups with quite different evolutionary histories. A few of them, including water lilies, belong to a group that branched early from the angiosperm tree. One of the earliest fossil flowers yet found is a tiny water lily dating from the Early Cretaceous, 125–115 million years ago. Monocotyledons ('monocots' for short) – which include plants with strap-like leaves and parallel leaf veins such as grasses, orchids and lilies (as distinct from water lilies) including daffodils, bluebells and tulips – have a genuine evolutionary affinity with each other and all belong to a group that branched off early too. About three-quarters of all flowering plants belong to a group of broad-leaved plants called the eudicots that evolved some time after the monocots diverged. Quite what made the eudicots so much more successful (in terms of numbers of species evolved) than anything that had come before is an interesting question that is currently being researched.

ABOVE Phylogeny of the flowering plants (angiosperms), showing some representative species.

RIGHT AND OPPOSITE
Three orchids showing their characteristically complex and colourful flowers.

Though the date at which the angiosperms split from some unknown gymnosperm ancestor is not precisely known, there is better evidence for later events. Much of the diversity in the eudicots, which comprise three-quarters of living angiosperms, evolved in the Cretaceous, between 142 and 65 million years ago, including the orchids, the most species-rich group among the monocots and the largest of all flowering plant families, containing at least 17,500 species. Orchids display to perfection the most significant feature of the angiosperms – their seemingly unlimited evolutionary potential to modify the flower.

Flowers vary extravagantly in shape, size, colour and odour and attract as pollinators an equally rich variety of animals. Practically anything that moves rapidly, especially by flying – bats, birds, beetles, bumblebees, other bees and butterflies (to name only the b's) – has been recruited by one flower or another to transport pollen, while the oblivious visitor flits from plant to plant in search of rewards such as nectar or even, as in the case of bee orchids, the illusory promise of a mate.

MADAGASCAN STAR ORCHID

Many flowers are unspecialized and can be pollinated by a range of insect species, but the orchids are especially particular about their pollinators, and this may be a key to their diversity. Observing that the nectaries in the flowers of the Madagascan star orchid *Angraecum sesquipedale* were located at the bottom of an 28 cm (11 in) long tube, Charles Darwin famously predicted that it must be pollinated by a moth with a proboscis of this length. It was 40 years, and Darwin was long dead, before the moth with its improbably long proboscis was actually discovered. It was a subspecies of an African moth *Xanthopan morganii* and was given the name *Xanthopan morganii praedicta* in honor of Darwin's prediction.

TOP RIGHT The Madagascan star orchid, *Angraecum sesquipedale*. Its nectaries are at the bottom of a flower tube that is nearly a foot long.

LEFT The moth with a very long proboscis that pollinates the Madagascan star orchid.

There can be little doubt that the capacity for flowers to evolve is key to the astonishing diversity of angiosperms. A new species can only arise when some barrier prevents a population from breeding with its progenitors, thus isolating it genetically (Chapter 11 *New species from old*). Therefore, anything that selects for evolutionary change in the characteristics of a flower, such as a change in a pollinator, may simultaneously create a reproductive barrier, creating the conditions necessary for a new species of plant to evolve. You might say that what was most significant about the initial evolution of the flower was its own evolvability.

ABOVE Flowers of *Mimulus lewisii* (a, b) and *M. cardinalis* (c, d) are pollinated by bumblebees and hummingbirds, respectively. The flowers on the left (a, c) are the normal colours for the species and those on the right (b, d) are the same species with their colours switched by cross-breeding. Switching colours caused pollinators also to switch between species.

How easily changes in flowers may produce reproductive barriers by altering the pollinators that visit them is illustrated by experiments with a pair of North American monkey-flower species (*Mimulus* spp.). A pink-flowered species, *Mimulus lewisii*, is visited by bumblebees while a closely related red-flowered species, *M. cardinalis*, is pollinated by hummingbirds. The difference in flower colour between the two monkey-flowers is controlled by a single gene which experimenters switched between the two species by cross-breeding. When the flower colours were switched, each plant attracted the pollinator that normally visited the other, proving that reproductive isolation between the species was determined by a difference in just one gene. Similar experimental results have been obtained with wild petunias.

Many of the insect groups that pollinate flowers diversified during the Cretaceous, apparently in parallel with flower evolution. Plants are prey to many insects as well as being pollinated by some of them. In many butterflies and moths, the adults are pollinators while the caterpillars are plant-eaters. The leaf beetles are a group of specialist plant-eaters containing more than 38,000 species. For every plant, there are tens-to-hundreds of insects and larger herbivores, including browsing mammals, that eat them and for each of these, there are numerous parasitic and predatory insects that attack the plant-eaters. Thus the rise of the angiosperms created an ecological feast upon which insect diversity gorged and multiplied. And, of course, one day humans would also exploit this diversity of plants.

LEFT Mint leaf beetle, *Chrysolina menthastri*. Dietary specialization contributes to the enormous diversity of beetle species.

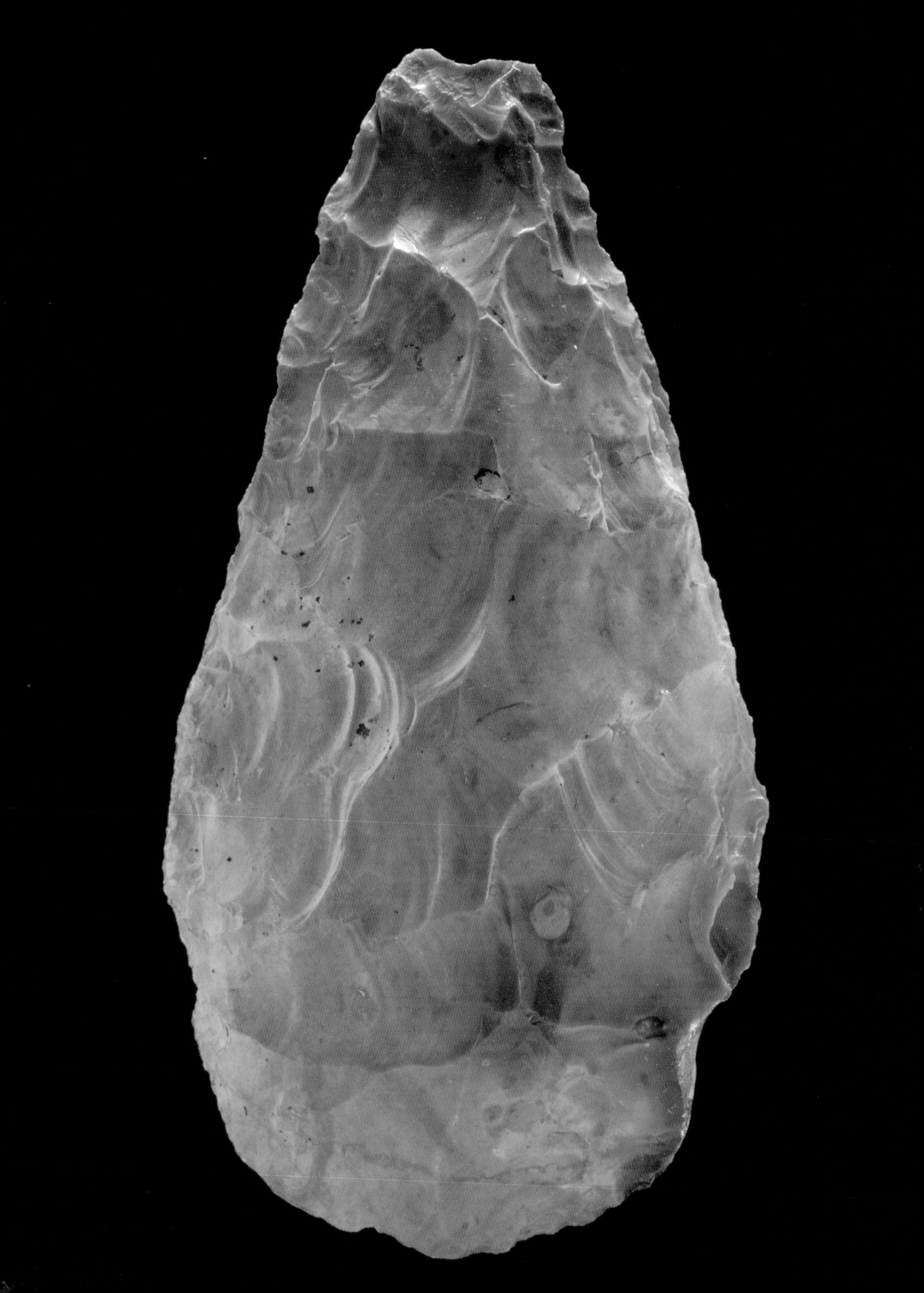

14 CHAPTER The race from Africa

IN FORMULATING THE HYPOTHESIS OF AN AFRICAN origin for humans, Darwin had no fossil evidence to guide him, as you read in Chapter 5 *African genesis*. After 150 years of research there are many fossil specimens and we have a picture of human evolution that is more advanced than his. Yet, it is still a very imperfect picture, particularly when set alongside the advances in understanding of the mechanisms of inheritance over the same period. There were once several species of human, but now there is only one. There were many species of other apes in the past too, but most are now extinct and those few remaining are endangered. There is no smoking gun, but given the devastating effect of *Homo sapiens* upon the planet, the sorry fate of our nearest relatives is likely to have been the consequence of our own spectacular success as a species. At what point in evolutionary history did the human apes of Africa spread out to colonize much of the globe? Was it a single exodus or a sequence of migrations, and what happened to the other species of *Homo*?

OPPOSITE Hand axes, like this British one, were made using Acheulean technology, first appearing in Africa and later spreading through the human populations. A suitable stone was chipped to produce flakes and the core stone was shaped into a sharpened tool, unlike the earlier Oldowan technology where the flakes themselves were used as cutting tools. This axe is 30.6 cm (12 in) long.

BELOW A cast of the 'Black Skull', *Paranthropus aethiopicus*, discovered near Lake Turkana in northern Kenya. The branch of the human tree that this specimen represents, the Australopithecines, became extinct over 1 million years ago.

The end of the line for Homo

Our species is the last survivor of our genus. Several species have existed since the human line first separated from the chimps. It is difficult to determine how many, but all except *Homo sapiens* are extinct. We are dependant upon fossil specimens for our knowledge of extinct species and fossils are exceedingly rare. When they are found, they are skeletal remains only and, to date, have almost always been incomplete. So, it is difficult to assign fossils to particular species.

The largest number of species that could be justified by anthropologists, at present, is 22, but there are some arguments from experts for reducing this number substantially, in an extreme case to as low as four. There are only small numbers of specimens, often from different populations and it is difficult to tell whether differences between specimens are a consequence of variation *within* a species or of variation *between* species.

In the period following the appearance of the first human – Toumai (Chapter 5 *African genesis*) or a similar species – the lineage became a bushy tree with a number of branches. Between 3 million years ago and 2 million years ago there was a group of robust-bodied species that left no descendants, the first representatives of our own genus, *Homo*, and a couple of side

The human saga written in DNA

RIGHT Some key archaeological sites in Africa that have provided evidence of technological and cultural advances between 75,000 and 55,000 years ago.

The idea of using mitochondrial DNA (mtDNA) to provide dates was introduced in Chapter 5 *African genesis*. The first big study of the mtDNA of living humans in 1987 gave a picture that appeared at the time to broadly match the fossil evidence. It showed that the mtDNA of Africans had slightly more mutations, when compared with other humans in the rest of the world, indicating that the African lineage was a little older. The study also showed that there were only a small number of differences between the mtDNA of all people sampled, implying a short period of evolutionary history since the last common ancestor of *Homo sapiens*. There is more genetic diversity in gorillas than there is in the modern human population, which not only indicates a relatively recent origin for the last common ancestor of our own species, but also suggests that there was a small initial population of *Homo sapiens* with a limited amount of genetic variation. A more recent study places the age of the last common ancestor at 143,000 years ± 18,000. This date does not perfectly align with dates from the fossil evidence, with finds of *Homo sapiens* from Herto and Omo Kibish in Ethiopia being dated at up to 196,000 and 150,000 years old, respectively. If these dates are accurate, then something very strange has occurred, for our species is about 200,000 years old but did not start to spread out of Africa and across the world until 65,000 years ago, over 100,000 years after splitting from the last common ancestor.

A further observation can be made from mtDNA studies. Periods of population expansion or contraction leave their mark on DNA. Comparing the same section of DNA from different individuals shows a number of mismatches in the code. The frequency distribution of mismatches in the code between individuals can be correlated with periods of population change. High frequencies of mismatch indicate population expansion. The evidence from the mtDNA of modern humans from different geographical regions shows that there was a clearly defined rapid expansion in the size of the African populations 80,000 years ago, followed by rapid expansions in Asia and Europe 60,000 and 40,000 years ago. So it appears that a rapid rise in the African population was followed by dispersal out of Africa and subsequent increases in the size of populations in Asia and then Europe.

Technological advances

RIGHT Two possible routes that *Homo sapiens* could have taken out of Africa and into Asia, about 65,000 years ago, one through North Africa and the other across the mouth of the Red Sea.

Archaeological evidence from South African sites such as the Blombos Cave and the Klasies River shows that there were a number of technological and cultural advances in the period 75,000 to 55,000 years ago. The advances include new types of blades and tools that appear to be designed for working bone and wood. Shaped bones have been found, along with sea shells that have holes in them, presumably because they were used as beads. There are two examples of incised red ochre which are the earliest definite forms of abstract art recorded. The collection of artefacts bears a striking resemblance to artefacts found in sites in Europe and West Asia, but crucially these sites are dated some 20,000 years later than the South African ones. Technology thus

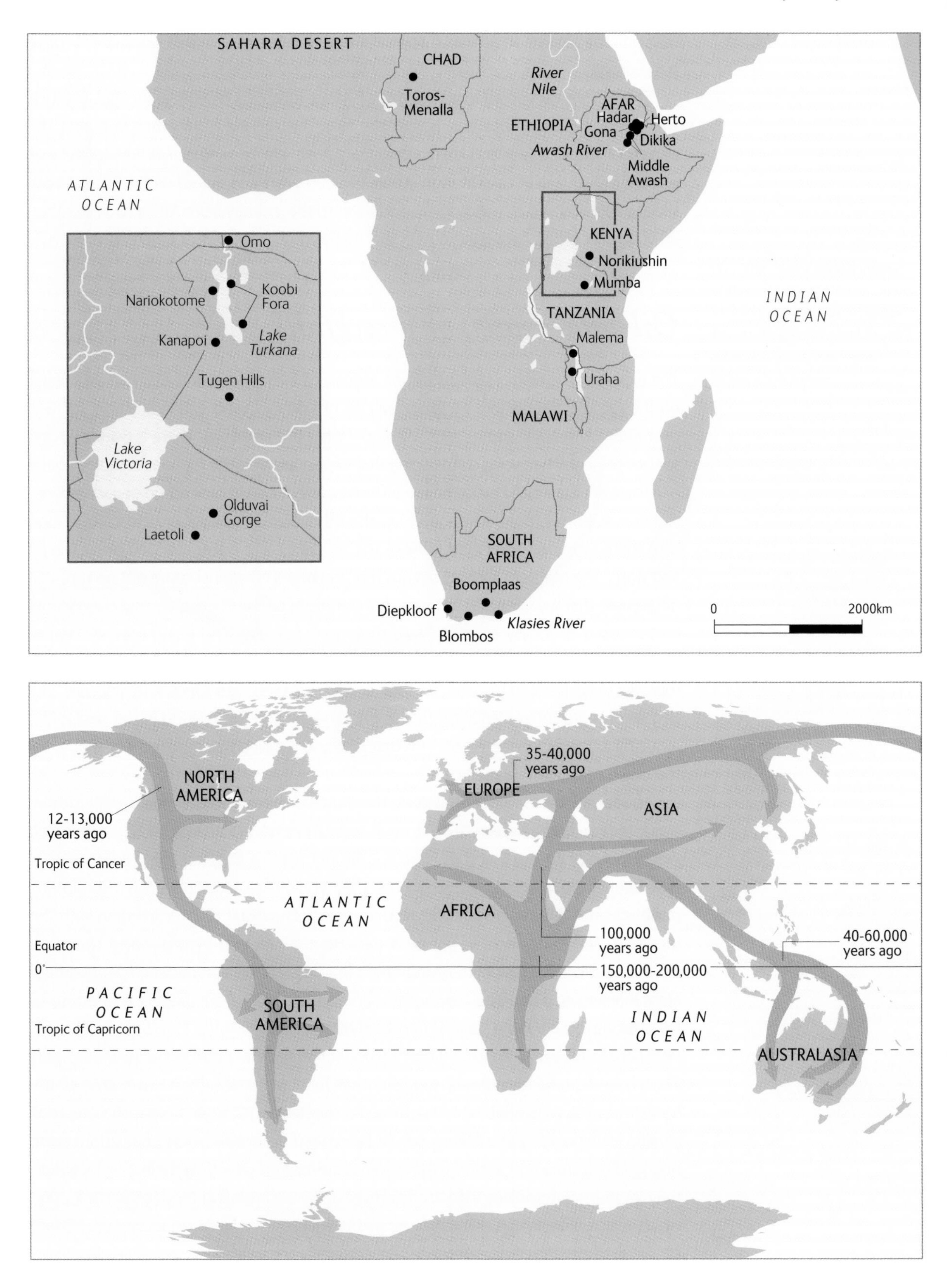
SAHARA DESERT
CHAD
Toros-Menalla
River Nile
AFAR
Hadar
Herto
ETHIOPIA
Gona
Dikika
Awash River
Middle Awash
ATLANTIC OCEAN
KENYA
Norikiushin
Mumba
INDIAN OCEAN
TANZANIA
Malema
Uraha
MALAWI
Omo
Koobi Fora
Nariokotome
Lake Turkana
Kanapoi
Tugen Hills
Lake Victoria
Olduvai Gorge
Laetoli
SOUTH AFRICA
Boomplaas
Diepkloof
Klasies River
Blombos
0
2000km
35-40,000 years ago
NORTH AMERICA
EUROPE
ASIA
12-13,000 years ago
Tropic of Cancer
ATLANTIC OCEAN
AFRICA
100,000 years ago
40-60,000 years ago
Equator
0°
150,000-200,000 years ago
PACIFIC OCEAN
SOUTH AMERICA
INDIAN OCEAN
Tropic of Capricorn
AUSTRALASIA

An analysis of Y chromosomes from a large sample of Africans and non-Africans showed that a particular mutation called M168, had its origin in Africa. The mutation is around 44,000 years old. Three other more recent mutations were identified in 163 Asian populations, all at the same site on the Y chromosome and thus there are three subgroups or polymorphic forms of the original M168 mutation, all ultimately tracing back to Africa. All the Asian men studied had one of these three polymorphic forms of M168, giving them a marker of their recent African ancestry. This strongly suggests that the replacement model is correct. If a man had been found without any of these mutations then it would have suggested an ancient origin for his Y chromosome that was outside Africa, but none of the 12,127 men tested lacked the Africa marker. This evidence supports the replacement theory for the colonization of Asia.

Neanderthals – a skeleton in our cupboard?

There is an ongoing project to sequence the genome of Neanderthals, a species of human that first appeared in Europe 175,000 years ago and became extinct about 27,000 years ago. They overlapped both in time and location with *Homo sapiens.* A short sequence of mtDNA, about a million base pairs, was obtained from a leg bone of a Neanderthal and this sequence was compared with the equivalent sequence from 994 modern humans. The mean number of differences between all the human sequences was 8 ± 3.1. A similar comparison between human and Neanderthal produced a mean of 27 ± 2.2. These data come from a short sequence only, but do suggest that the Neanderthal sequence is outside the range of modern humans and that, as a consequence, the last common ancestor of modern humans *and* Neanderthals lived well before the last common ancestor of all modern humans. Neanderthals therefore are not our direct ancestors. The replacement theory is therefore supported by the first results from the sequencing of Neanderthal DNA.

The race from Africa, timed by the molecular clock

The majority view among scientists is that *Homo sapiens* originated in Africa and, about 65,000 years ago, spread across Asia and then into Europe and Australasia, replacing the other species of humans then living, such as *Homo erectus* and *Homo floresiensis* in Asia, and *Homo neanderthalensis* in Europe. They would argue that there is a consistent picture given by the fossils, the dating methods and the DNA analysis, all of which provide evidence for the race from Africa, but there is no consensus view. There are strong, vocal proponents of the multi-regional hypothesis and they are very critical of the evidence from DNA, principally because of the assumptions about mutation rates that are crucial to dating events using the molecular clock (Chapter 5 *African genesis*).

The debate between the majority 'Out of Africa' proponents and the minority multi-regionalists is acerbic, heated and occasionally personal. It resembles the debate from entrenched positions that surrounded the publication of *On the Origin of Species* and, just like that debate, it is stimulating new research. Debate based on evidence, observation and experiment is a fundamental part of the progress of understanding in science today, just as it was in Darwin's time. That we are still seeing debate about the African origin of *Homo sapiens* 150 years on is indicative of the difficulty of assembling evidence of our fossil ancestry, but the search for a definitive explanation for the origin of *Homo sapiens* remains as attractive a quest as ever.

15 CHAPTER Catch me if you can

SINCE ANCIENT TIMES, DISEASES WERE ASCRIBED many different causes, from the alignment of planets to bad air (hence 'malaria'). Although from the time people lived in cities it was clear that some diseases were contagious, the concept of 'germs' that were living pathogenic organisms was not firmly established until late in the 19th century. Germs are very diverse; most are bacteria or viruses, but some are single eukaryotic cells, fungi or multicellular animal parasites, and can cause disease in all kinds of organisms, including each other. Some pathogens pass directly from host to host, but others are transmitted by vectors, often insects or other highly mobile animals.

BELOW Dutch elm disease is caused by fungi spread by small beetles that bore tunnels in and under the bark. The tree's reaction to the fungus blocks its water-carrying vessels, killing the tree from the top downwards.

Many pathogens can proliferate rapidly in favourable conditions, and so, like HIV (Chapter 1 *99% Ape*), can evolve quickly. Fortunately, our natural defence against disease, the immune system, can produce such a wide range of defensive proteins that it is rarely outwitted by new forms of pathogen. Over the longer term though, pathogens and hosts are engaged in an evolutionary arms race that can end in truce or disaster.

Truce or consequences

From time to time and place to place, large numbers of people, animals and plants die quickly from epidemics of diseases. Dutch Elm disease appeared in Britain in the late 1960s and spread rapidly, so by 1980, mature elms, previously among the commonest native trees, became rare.

In 1988, thousands of seals around northwest Europe died from a viral disease. Many people, including previously healthy adults in their prime, have died from cholera, smallpox, plague, measles, influenza and other contagious diseases. These outbreaks almost always disappear as rapidly as they arise and, by themselves, seldom exterminate an entire population because both the pathogens and the populations they infect are evolving.

OPPOSITE A mature elm, *Ulmus*, dying of Dutch elm disease. Elms are long-lived trees; healthy specimens can resist most of their pathogens but succumb to newly-introduced diseases.

FRIEND AND FOE

Most pathogens are viruses, bacteria or (especially in plants) fungi, but most such microbes are not pathogens. All animals (especially, but not exclusively, eaters of tough plants) harbour billions of benign microbes in their guts as aids to digestion. If these little helpers are killed, for example, by administering powerful antibiotics, their hosts starve because they cannot digest their food adequately. There are no consistent differences between these collaborators and pathogenic microbes, certainly none that could be identified by simple microscopy.

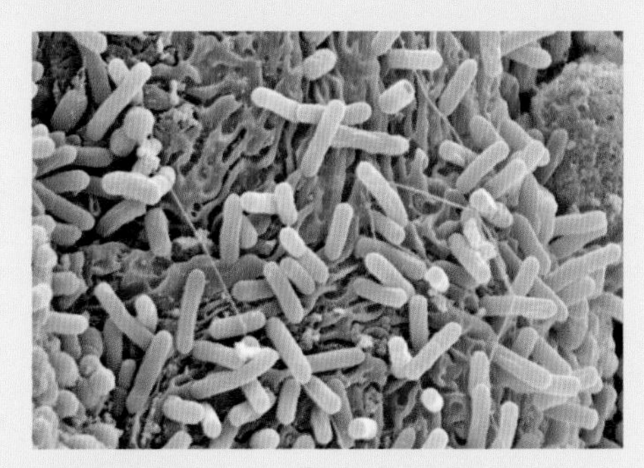

Some species of normally harmless bacteria occasionally become pathogens. A familiar example is *Escherichia coli*, better known as *E. coli*, of which billions are found in almost everyone's faeces (and those of other mammals as long as they are not hibernating – *E. coli* likes to be warm), where they help break down the solid waste and promote the uptake of vitamin K and other important nutrients. But if the bacteria escape from the gut (for example, through a bleeding ulcer or wound) and proliferate in the blood or elsewhere, or those normally found in other species are eaten with contaminated food, they can cause fatal disease, including severe diarrhoea, meningitis, peritonitis, septicaemia and forms of pneumonia.

ABOVE *E. coli* (about 1/1,000 mm long), shown as pale brown rods by scanning electron microscopy, living harmlessly in the gut.

Evolving disease

Even distantly related species share a surprisingly high proportion of their genes. Consequently, many microbes can live and proliferate in many different hosts. Most human pathogens can infect chimpanzees, gorillas and other primates, though their presence may not cause disease. Wild apes on tourists' itineraries and those in zoos can catch colds and influenza from their visitors. People also share microbes, some pathogenic, some harmless, with more distantly related animals with which they frequently come into contact, including dogs, pigs, cattle, rats and domesticated poultry.

Because they can proliferate so rapidly, microbial populations recover quickly from strong natural selection. As long as they are able to breed, disease-causing microbes can evolve very rapidly. Many, like HIV, have genetic material that is especially prone to mutations. But equally important as sources of variation are genes acquired from other pathogens, transmitted from one species to another in minute viral particles.

Gene swapping

When first used in humans, antibiotics such as penicillin were miraculously efficient, eliminating many kinds of bacteria (useful and pathogenic) in a few days. Penicillin works by preventing the formation of the organisms' protective coat; other antibiotics, such as tetracycline, interfere with protein synthesis or gene copying. Within a few years of being widely used, each new antibiotic was reported to be ineffective in certain cases.

Strains of bacteria could proliferate even in the presence of large quantities of antibiotic. These mutants often breed quite slowly and cannot compete with normal bacteria under favourable conditions, but in the presence of the antibiotic, all the sensitive bacteria die, leaving the antibiotic-resistant ones to proliferate unhindered: natural selection favours mutant bacteria that are better adapted to the new conditions. The more often bacteria and antibiotics come into contact, the more likely this sequence of rare mutation and strong selection is to occur.

At least four very different mechanisms of antibiotic resistance have evolved during the past half-century. The genes involved have passed between different species of bacteria in viral particles so efficiently that some human pathogens, including those that cause tuberculosis, now have the genes for all four. Antibiotics are almost useless against such multiple drug resistance.

Ideal homes for microbes

The host's death is always bad for pathogens, especially those most specialized to life as a parasite – they have lost their home and the opportunity to spread to other hosts. Most outbreaks of lethal diseases, such as plague, typhus and epidemic influenza, disappear after about two years because the pathogens kill too many of their hosts too quickly. Unless people are living very close together continuously, as in cities, ships and armies, each infected person does not live long enough to transmit the pathogens to enough other people to sustain the epidemic. In this respect, HIV has done much better; the long delays before symptoms appear enable its hosts to infect many other people before succumbing to AIDS.

Pathogens that normally cause serious illness occasionally establish themselves in people sufficiently for them to remain infectious without showing symptoms of the disease. A famous case was that of an Irish cook working in New York City during the first two decades of the 20th century. Health officials investigating outbreaks of typhoid found that the disease appeared in households shortly after Mary Mallon (1869–1938) began work as a cook. Mary herself never became ill and refused to accept that she could be the source of the infection that devastated her employers' families. Eventually, the city authorities incarcerated her in an isolated sanatorium. But several years later, Mary won her freedom on the condition that she never again worked as a cook. After her release, she adopted a pseudonym to evade this restriction, took another kitchen job and infected numerous patients at a maternity hospital. 'Typhoid Mary' was rearrested and confined to an island outside New York, where she died of pneumonia 23 years later.

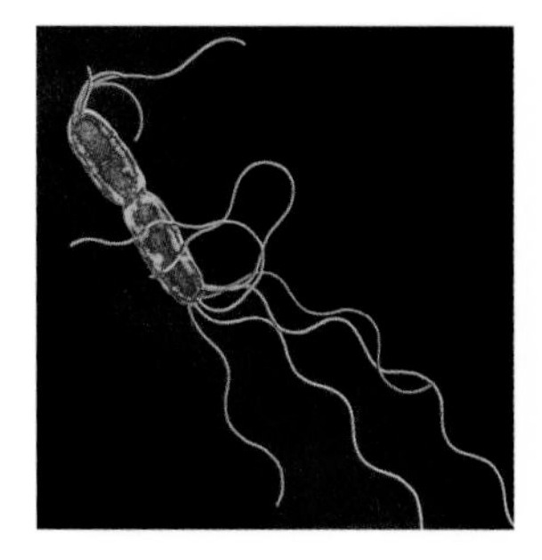

ABOVE False colour image of dividing *Salmonella typhi*. These 1/1,000 mm-long bacteria have several whip-like flagella and can swim in warm, stagnant water as well as inside guts.

Because Mary's physiology, for reasons that are still not properly understood, tolerated the typhoid bacteria, she disseminated them far more widely and infected many more people than she would have done had she become ill and either died or exterminated them from her gut. About 5% of people continue to harbour the bacteria after they have completely recovered from the disease, usually for only a month or two, but a few remain typhoid 'carriers' for years. Such people are important to the evolution of the

drunk by another potential host. Most strains of cholera cannot survive outside warm bodies for longer than a few hours, so transmission is only efficient in cities, armies and other dense populations. Modern sanitation, frequent hand washing and, probably most important of all, tea-drinking (for which the water is boiled) have all but eliminated opportunities for transmission of cholera bacteria in Western conurbations, but the disease was common in London and other large European cities until the early 20th century. Outbreaks still occur in countries where sanitation is less efficient and human waste contaminates drinking water.

Some pathogens are unable to survive even briefly outside their host and hence can be transmitted only by direct physical contact and the transfer of body fluids: blood, saliva or semen. Examples include most of the sexually transmitted pathogens, including HIV, the malarial parasite (which passes part of its life-cycle in a mosquito and part in a mammal), Lyme disease (transmitted by biting ticks to humans and many other mammals, especially deer), and rabies (usually transmitted by an infected animal biting another).

Pathogens may evolve the ability to survive outside the body or to dispense with vector organisms. Tuberculosis was, and still is, a disease of cattle. People probably only caught it by prolonged, intimate contact with infected cattle or by drinking raw milk from diseased cows, but several millennia ago, possibly in ancient Egypt, the pathogens evolved the ability to pass directly from person to person as airborne particles, and thus the pathogen has become much more widespread. HIV is transmitted from person to person only by sexual intercourse or by blood transfusions because it cannot survive outside its host. What would happen if it acquired the ability to form airborne particles, as the viruses that cause common colds have done?

BELOW Insects living as parasites on mammals are intricately adapted to their hosts' structure and habits. Not all transmit pathogens, and some cause little irritation. This scanning electron micrograph shows species-specific features of legs, mouthparts and body.

Lousy evolution

Most mammals harbour lice and fleas; they live in the pelt, eating flakes of skin or sucking blood (or both), passing from mother to young during suckling and between adults during fighting or mating. Because mammals have different kinds of hair and grooming habits, most have their own unique species of flea or louse. Hedgehog fleas are among the largest and most conspicuous: spines are cooler than fur and are too long to be groomed by short-legged hosts specialized for rolling up into a ball. Most other mammalian fleas are smaller and more agile, adept at evading vigorous grooming.

Many human diseases, including some of the most ancient that also afflict other primates, spend part of their life-cycles in blood-sucking animals such as lice, fleas, ticks or mosquitoes, which transmit them from host to host. Fleas cannot feed from dead mammals – the tiny superficial blood vessels that they puncture collapse as soon as the heart stops – so they quickly move onto the nearest warm food source. Rat fleas also bite humans and can transmit the pathogens that cause plague between the two mammals; both can die within hours of developing this virulent disease.

Comparative studies of DNA show that speciation of hosts and their lice occurs almost simultaneously. Most mammals harbour just one louse species but we have three. The body louse (*Pediculus humanus humanus*) lives in clothing, not hair; it can withstand cold and up to a week without feeding, conditions that would kill most other lice. When warm, lice crawl on skin, causing intense irritation. The accumulation of neutral mutations in louse DNA suggest that we created the new habitat to which body lice are adapted 100,000–70,000 years ago, as humans

ABOVE Fleas are laterally flattened, wingless insects usually 1-4 mm long. The adults are bloodsuckers, which reach new hosts by jumping up to 200 times their own body-length using catapult-like mechanisms in their long hindlegs.

RIGHT Lice are dorso-ventrally flattened, wingless insects that gnaw skin. The human pubic or crab louse, *Pthirus pubis* , about 2 mm long, resembles lice found on gorillas, and recent research shows that their genes are also similar.

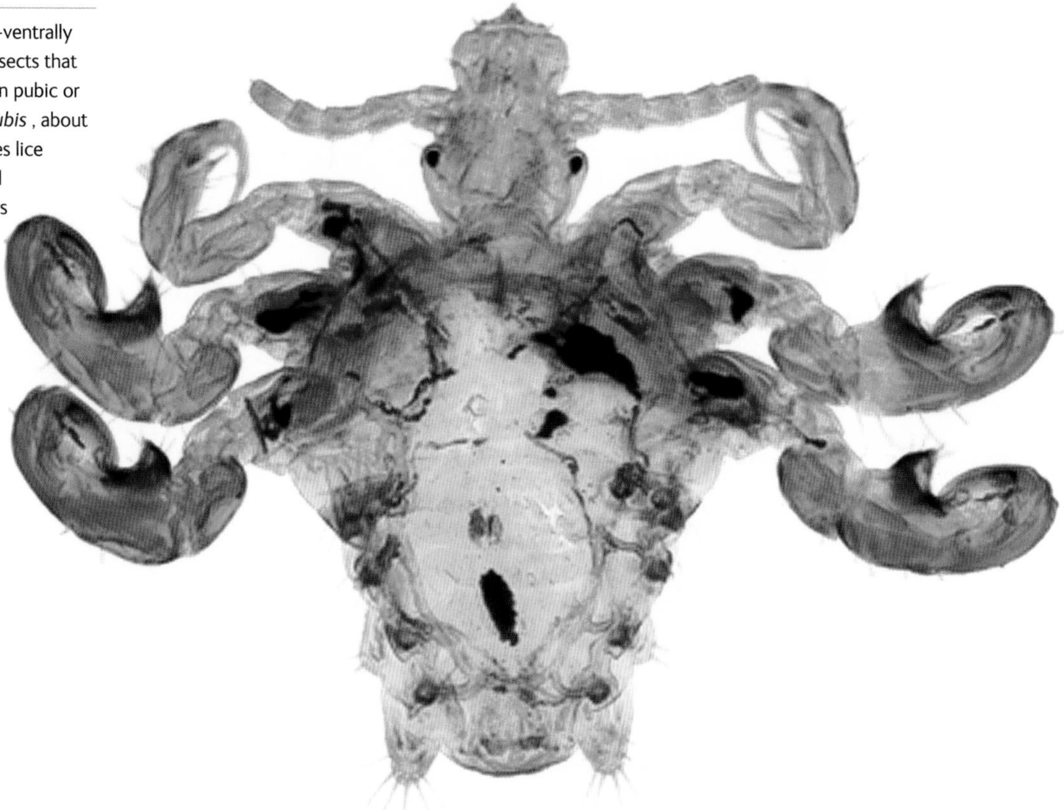

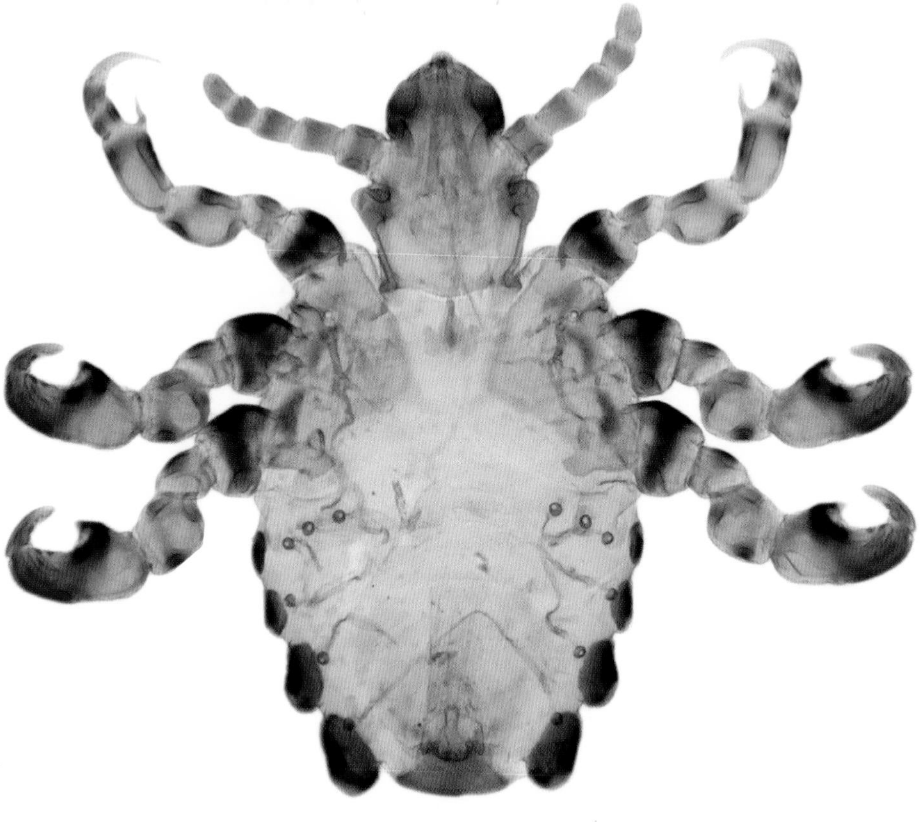

RIGHT *Pthirus gorillae* is found on gorillas. The hooked feet are adapted to grasp their hosts' hairs, enabling the lice to resist scratching and grooming.

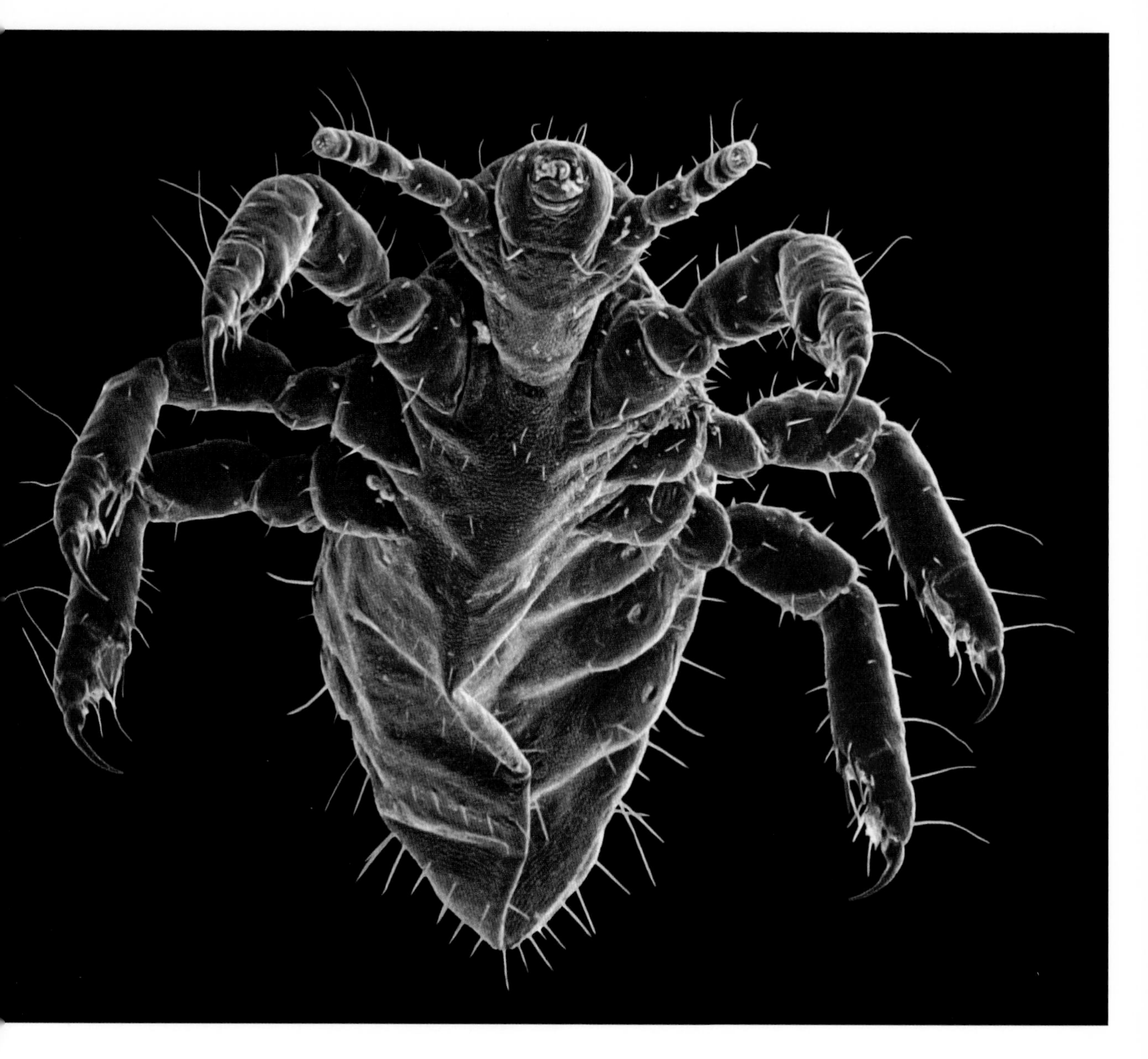

migrated from Africa into the cooler climates of Europe and Asia. The bacteria that cause the most dangerous form of typhus fever in humans also proliferate in lice. Although infected lice become ill and die, in crowded conditions they may still live long enough to transmit the pathogen to more people.

Dry cleaning and washing machines have hit body lice hard (and so almost eliminated typhus in the West), but a subspecies, head lice *Pediculus humanus capitis*, survive hair washing and combing. Lice cannot fly but they run fast, so head lice, which can also harbour typhus, readily pass between children in nurseries and schools and are now common in Britain. *Pediculus humanus* is a sister species to chimpanzee lice, but the smaller, rounder pubic or crab lice (*Pthirus pubis*) are most closely related to the gorilla louse. Its ancestors may have migrated from gorillas to humans while the two species were living in the same areas of Africa about 3.3 million years ago.

ABOVE The adult human head louse, *Pediculus humanus capitis*, here seen from below, is up to 4 mm long. It is longer and slimmer than *Pthirus*. In this scanning electron micrograph, the abdomen has shrunk and appears twisted.

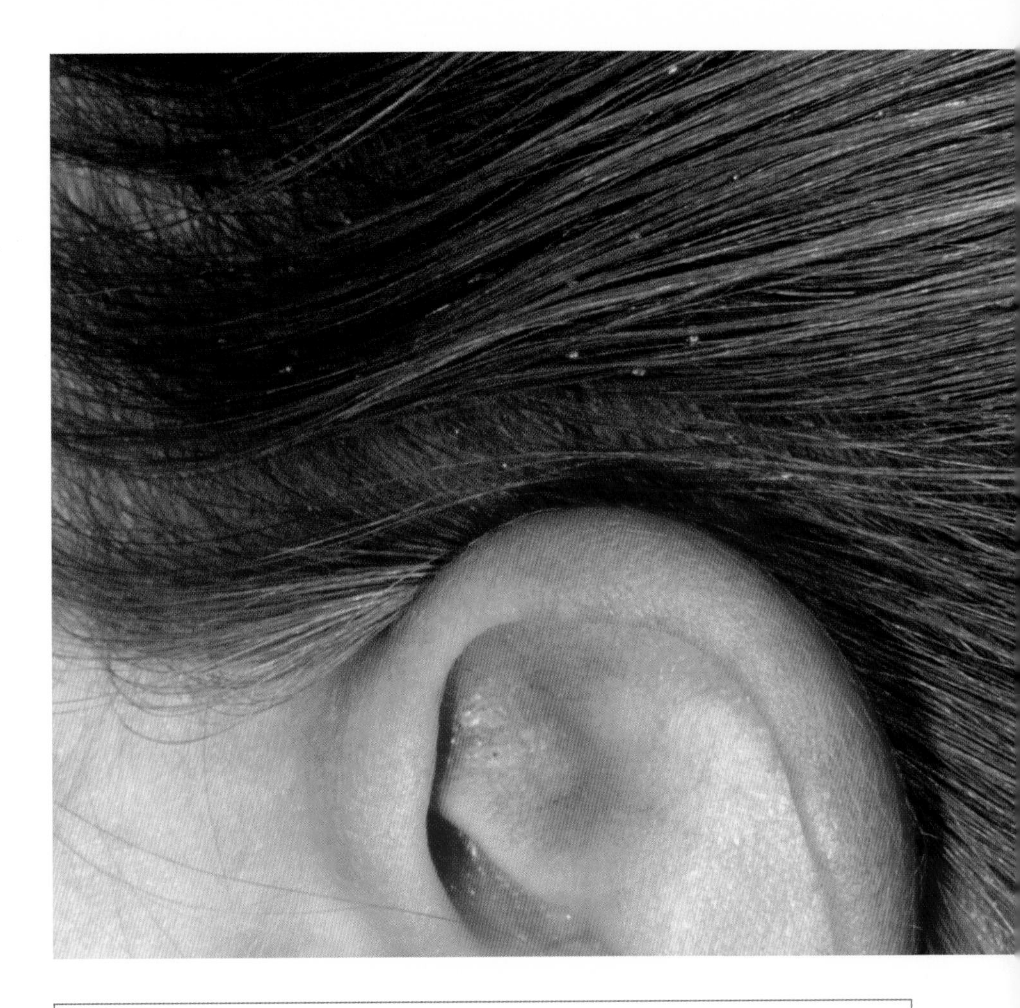

RIGHT Louse eggs called nits stick to hairs and are remarkably tough, surviving hair brushing and most shampoos.

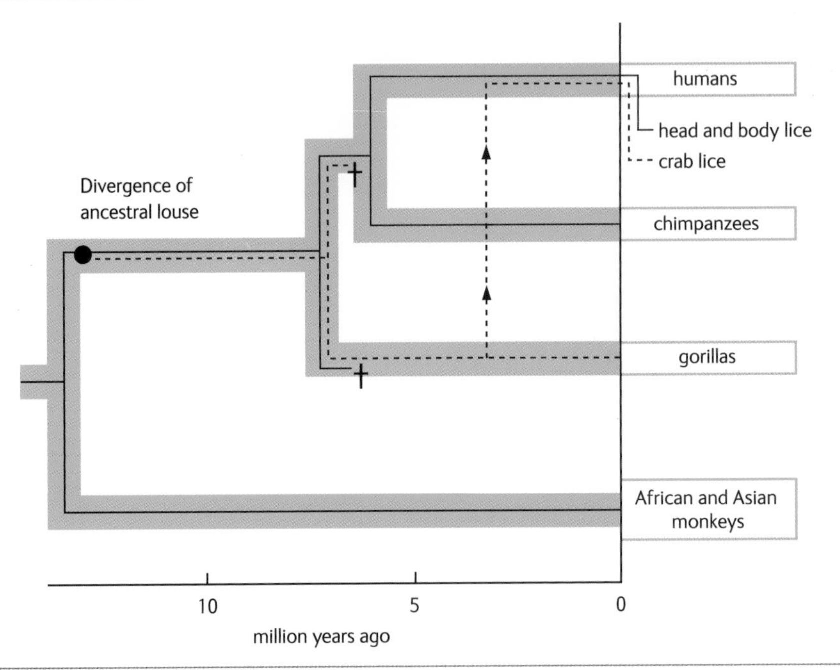

RIGHT Cladograms showing the relationships between the two genera of human lice and those of other apes, and between the primates themselves.

Pathogens provide some of the best examples of evolution in action owing to their short generation time, labile genes and adaptations to highly specialized ways of life. They evolve ways of coexisting with their hosts, upon which they depend, of surviving the hosts' attempts to exterminate them and of colonizing new hosts more efficiently. Modern humans offer opportunities for many pathogens, with their long lifespan, prolonged childhood, living in permanent, high-density settlements in close association with dogs, rats and other livestock, taste for long-distance travel and generally promiscuous sexual habits. These topics bring us to a consideration of human society, and in particular, how our minds are adapted to function in social situations.

16 CHAPTER Darwin in mind

IN THE CLOSING CHAPTER OF *ON THE ORIGIN OF SPECIES*, Darwin made the prediction that "Psychology will be based on a new foundation". He did not elaborate on this statement, but thinking about the mind in the same way as we now think about the function of other organs is still controversial. To understand the human heart, it is extremely helpful to have some idea of its role in the functioning of the whole organism. That is, the study of how it works, and how it grows, is illuminated by understanding that it is a pump that circulates blood which supplies oxygen and other essentials to body tissues. Various features of the heart – such as its having four chambers, their relative proportions, and the timings of contractions of its muscular pump – can be understood as design features on which natural selection has acted. The function of the heart in the body may be deduced from how its structure has been shaped by natural selection. It is curious that we do not usually think of the mind in the same way, because the mind has been subject to very strong selection pressures throughout our history.

Evolutionary psychology

Your mind is not some neutral by-product or vestige like your earlobes or the whorls on your fingers. It is your mind that you use to detect and avoid a danger. It is your mind that you use to decide when and what to eat. It is your mind that you use to attempt to select a mate and decide when to reproduce. It is your mind that you use to try to manage the complex give-and-take of social relationships with other individuals. It follows that the kind of mental structures and decision-rules that your ancestors possessed would have been a huge determinant of their reproductive success, and thus that natural selection has honed a suite of mental adaptations for solving particular recurrent problems of living.

The Darwinian position outlined in the preceding paragraph is that taken in the approach to the mind called evolutionary psychology. Why the basic premise of evolutionary psychology should still be controversial is an interesting question. One answer is that people misunderstand what it entails, an issue we return to later in this chapter. A deeper answer comes from evolutionary psychology itself. It seems that our evolved mentality divides the world into human agents, whose behaviour is to be thought of in terms of desires, beliefs and so on, and physical stuff, which is thought of in terms of causal processes with no intentional component. We evolved this dualistic way of thinking because it works pretty well in everyday life, but it becomes misleading when we try to do science. Neuroscience and evolutionary psychology consider the mind as

OPPOSITE The mind is an evolved organ just as much as the heart is, but people find this difficult to accept.

the outcome of various types of causal process with no intentional component. As a consequence, many people resist their implications. For example, most children believe that their brains are not involved in loving their siblings or brushing their teeth. People often say that they chose their spouse because they loved him or her, not because they were trying to maximize their reproductive success. Obviously, though, these explanations are just at different levels. It is true people choose their spouse because they loved him or her, but the likely reason that they have a propensity to fall in love, and that love has a particular logic and set of triggers to it, is because over thousands of generations, ancestors who had this emotional package left more offspring than those who did not.

Although Darwin's theory has been around for 150 years now, evolutionary psychology has only really got going in the last two decades. The reasons for this long delay are several. Partly they relate to the more general delay in understanding the underlying genetics and behavioural implications of natural selection that followed Darwin's seminal work. Partly they are to do with the time it has taken to generate specific, testable predictions from the overall Darwinian theory in areas that psychologists are concerned with, such as emotion, parenting, aggression, relationship behaviour, and so on. And partly, it is sociological. Psychologists are not routinely trained in biology, and so may misapprehend the difference in level of explanation, and the various ways that exist of testing evolutionary hypotheses.

Evolutionary psychology does not seek to replace any of the existing theories of psychology (or at least, it should not). It merely adds a new focus. Psychology has traditionally been concerned with two questions. The first is that of *mechanism*. In

RIGHT Evolutionary psychology does not deny the importance of feelings like love. Instead, it asks the ultimate 'why' question of where our capacity to feel them comes from.

other words, what are the immediate processes in the mind or brain that give rise to a particular pattern of behaviour or experience? For example, you might be interested in knowing what types of sequence of thoughts lead to aggressive behaviour. The second is that of how a particular pattern of behaviour or experience develops over the course of the individual's life (biologists call this the question of *ontogeny*). For example, you might wish to know whether seeing aggressive behaviour in childhood makes people more likely to behave that way later on. Both of these questions are perfectly respectable. Evolutionary psychology adds to these the additional angle of *ultimate causation* – namely, how did it enhance the reproductive success of your ancestors to have the capacity to be aggressive, or the capacity to learn to be aggressive.

The question of ultimate causation is not just interesting in its own right. If we understand how aggression evolved, then many of its more immediate features may begin to make sense. The most common type of human aggression occurs between individuals of the same sex in defence of status or reputation. It is a risky strategy, in as much as it can lead to harm or social sanction, although in certain ecological contexts it can bring benefits. Since men have greater variance in reproductive success than women (that is, some men have no offspring but a few may have dozens – many more than any woman can have), we can predict that pursuing an aggressive strategy is more often adaptive for men than for women, especially while they are young and single, when reproductive competition is at its highest. These predictions turn out to be true. Across hundreds of studies in many cultures, men are more violent and aggressive than women, especially when they are young. When they marry, their testosterone levels decline and their likelihood of aggression does likewise (even controlling for age). If you are a man, you may notice that your car insurance premium is higher than that of your female peers, and declines steeply once you reach the age of about 35, or if you marry. Insurance actuaries analyse the age and status of those involved in accidents and thus understand exactly the evolved design features of risky and aggressive behaviours (if not their origins).

Note also that the evolutionary psychology perspective is nothing to do with claiming that aggression is 'genetically determined' or 'inevitable'. On the contrary, aggression is a high-cost, highly risky strategy, and it would be extremely maladaptive to just activate it in some indiscriminate way. Ancestors with such a tendency would not have fared well. Instead, selection has made us reluctant to deploy aggression unless we have little to lose. In modern societies, social deprivation, in terms of relative poverty and poor employment prospects, is a very strong predictor of aggression and violence. This makes perfect sense from the evolutionary psychological perspective. If you give people nothing to lose, then the evolved cost–benefit calculation that goes on in their minds is swayed towards the aggressive end of the spectrum. It follows that, even though the psychological mechanisms that cause aggression have been shaped by natural selection, the immediate triggers of aggression are environmental, and the occurrence of aggression is best altered by ameliorating the social environment.

Psychologists will only become convinced that the evolutionary perspective is useful if they can see that it helps them understand something they would not otherwise have understood. Fortunately, there are many examples of such breakthroughs. A recent one comes from the study of attention. Psychologists have been studying the processes of attention (basically, the way you devote more processing to some aspects of the environment than to others) for decades, but produced generally banal findings, such as the fact that you can change the way your attention is allocated by having a particular goal or task in mind, and that as you become expert in some domain you start to attend to different features of the stimulus than a novice would. But let's think about what attention is for. The major function is to identify things in your environment whose behaviour is going to be important to you so that you can devote more mental resources to monitoring them. What kinds of things would have been important for our ancestors?

Hunter–gatherer hangovers

This prediction has been tested using what psychologists call a 'change detection' method. It works a bit like the 'spot the difference' competition you sometimes find in the back of newspapers. People are presented with a picture twice in quick succession, with one component having been deleted or added in the second presentation. They have to detect the change as quickly as possible. The more attention the participants deploy to an area of the scene, the quicker and more reliably they detect any changes there.

For most of our history we have lived as foragers, in intensely social groups with a lot of cooperation and some conflict, mostly outdoors in unbuilt environments. The most important things to monitor in such contexts would be the other *people* (you might be cooperatively foraging with them, or involved in a stand-off with individuals from another group), and *animals* of other species (you might be hunting them, or, unfortunately for you, they might be hunting you). This simple observation leads us to a prediction: people should spontaneously devote more attention to humans and other animals in the visual environment than to other components of similar size and visual impact.

BELOW Your ancestors have spent most of the last hundred thousand years in visual environments very different from those that you live in now, where the key entities to be tracked were animals and other people.

The results are compelling. Participants detect changes to people or other animals in the scene more quickly and reliably than any other components. This difference can't be accounted for by the people and animals simply being more visible in the scenes used, since the images were matched for size and contrast. Moreover, if the pictures are turned upside down, so that recognition of what the objects are is abolished, then the higher detectability of images of animate objects disappears. The effect cannot be put down to exposure or current context. The

ABOVE People are better at detecting changes in complex images when the change is to an animal or person than when it is to an inanimate object, even a familiar moving one such as a car.

LEFT Cars are the most dangerous objects in our current environment, yet many more people are afraid of spiders or snakes than are afraid of cars.

participants were all young Americans who see motor vehicles thousands of times more often than they see elephants, yet they found elephants much more attention-grabbing than motor vehicles. Motor vehicles are also the most likely thing to kill them in the environment in which they live and yet (unfortunately for road safety) they are not what their attention systems have evolved to track.

Similar results have been found in other situations. People find snakes and spiders particularly attention-grabbing and frightening. These findings illustrate two features of evolutionary psychological research. First, there are influences of the ancestral

LEFT AND OPPOSITE Spiders and snakes hold a vestigial terror for many people, but the threat they actually pose in most countries is minimal.

environment, not just the contemporary one, on mental mechanisms. We fear the things that killed our ancestors often to a greater extent than the things that are likely to kill us. Even in countries where venomous spiders and deadly snakes occur, road traffic kills thousands of times more people a year, yet it is easier to find individuals with phobias of snakes and spiders than of motor vehicles. Second, although the explanation is in terms of the evolutionary past, the behaviour to be explained is that seen in the present. Critics often ignore the evolutionary perspective because they are more interested in what people do now than what happened in the Pleistocene. However, this objection is based on a misapprehension: these studies are about what we do now. It is just that what we do now is illuminated by understanding our evolutionary history.

Culture club

A common misunderstanding of evolutionary psychology is the claim that it implies culture is unimportant, and that the behaviour of humans in all societies is the same. Clearly, human thought and behaviour varies greatly with social and cultural context. Consider sexual attractiveness. Sexual attraction plausibly has an evolved function, namely to select an appropriate mate with whom to reproduce. However, norms of attractiveness vary. When young male Britons are asked which of a range of female body shapes they find most attractive, they prefer those that are neither extremely lean, nor extremely ample. Their optimum is for a body mass index (a measure of fatness adjusted for height, defined as weight in kilograms divided by height in metres, squared) of around 20, which is quite low, and at higher values of BMI, attractiveness tails off sharply. South

ABOVE Different cultures agree that very lean female bodies are not attractive, but they disagree about how much fat is good.

African male Zulus, by contrast, find a female body mass index of 40 just as attractive as one of 20. They agree with the Britons that too lean is unattractive, but for them, bigger is better, or at least, just as good.

One response to this dramatic cultural difference would be to say, 'Aha! It's not an evolved preference. It's culturally determined.' However, this reaction would be illogical, since culture is not the opposite of evolution but one of the ways evolution designs adaptations. Consider this. We are a species that has lived in many different environments with different nutritional bases and disease regimes. Thus, the kind of body shape that best indicated health and fertility would have been variable in space and time. It would be suboptimal for selection to build a fixed preference into us, and much better for it to build into us a general principle ('Pay attention to body shape as it is likely to be an important cue'), and a learning rule (something like 'Learn what is the most valued type of body shape among the other people around you' or 'Learn which body shape the healthiest people in your social group have'). This evolved mental rule then gives rise to the cultural difference. Thus, cultural variation is immediately explained by people's learning history, but more deeply explained by the pattern of underlying mental adaptations, which have learning rules built in. Such flexibility in preference is not unique to humans; many animals have evolved a high degree of flexibility in behaviour according to local ecological context. An ability to learn from others permit adaptation to local conditions without making expensive mistakes.

Evolutionary psychological reasoning predicts that cultural variations should be extremely widespread, since we are a generalist species that has made its living in dramatically different ways over the millennia, and has clearly been able to adjust to the demands of so doing. In fact, some of the most exciting parts of evolutionary psychological research concern cultural differences. For probably the first time, we have a theory based on principles of why such striking and patterned differences in behaviour appear among genetically indistinguishable humans living in different social and physical situations.

There is one final resistance that people have to evolutionary psychology: the idea that it diminishes free will. We don't like the thought that we are just blindly executing evolved drives or rules, and we prefer the thought that we have the autonomy to decide for ourselves. This problem is not in fact restricted to evolutionary psychology. It is a feature of any scientific account of mental processes, which is bound to rub up against our intuitive sense of free will and self-determination.

Explaining free will is a deep problem, but in fact evolutionary reasoning may have something useful to say about it. The world is a hazardous and uncertain place and raising offspring is an expensive investment for a human parent, requiring a long period of parental care. In some animals, and indeed plants, the hazards of an unpredictable environment have pushed the evolution of parental investment in a different direction, towards the production of vast numbers of offspring with relatively little investment in each. A few are then likely to make it through. Humans have gone to the opposite extreme. Parents can only produce a few children during their lifespan. A great deal of resource goes into making them, and they take a long time to grow. Part of the return on this investment is a very large brain, which an offspring can use to work out the best way to protect its fundamental interests and objectives in the current context in which it finds itself. In other words, our high degree of agency and autonomy – our ability to think and make up our minds for ourselves – is a legacy of the evolutionary history we have, and not its antithesis. Where other animals might have been selected to be fast, or light, or nimble, we have been selected to be able to work things out – and that, of course, is the ultimate explanation for why it took a human, Charles Darwin, to figure out where we have all come from. But even Darwin was unable to find a satisfactory solution to another evolutionary puzzle: why be good?

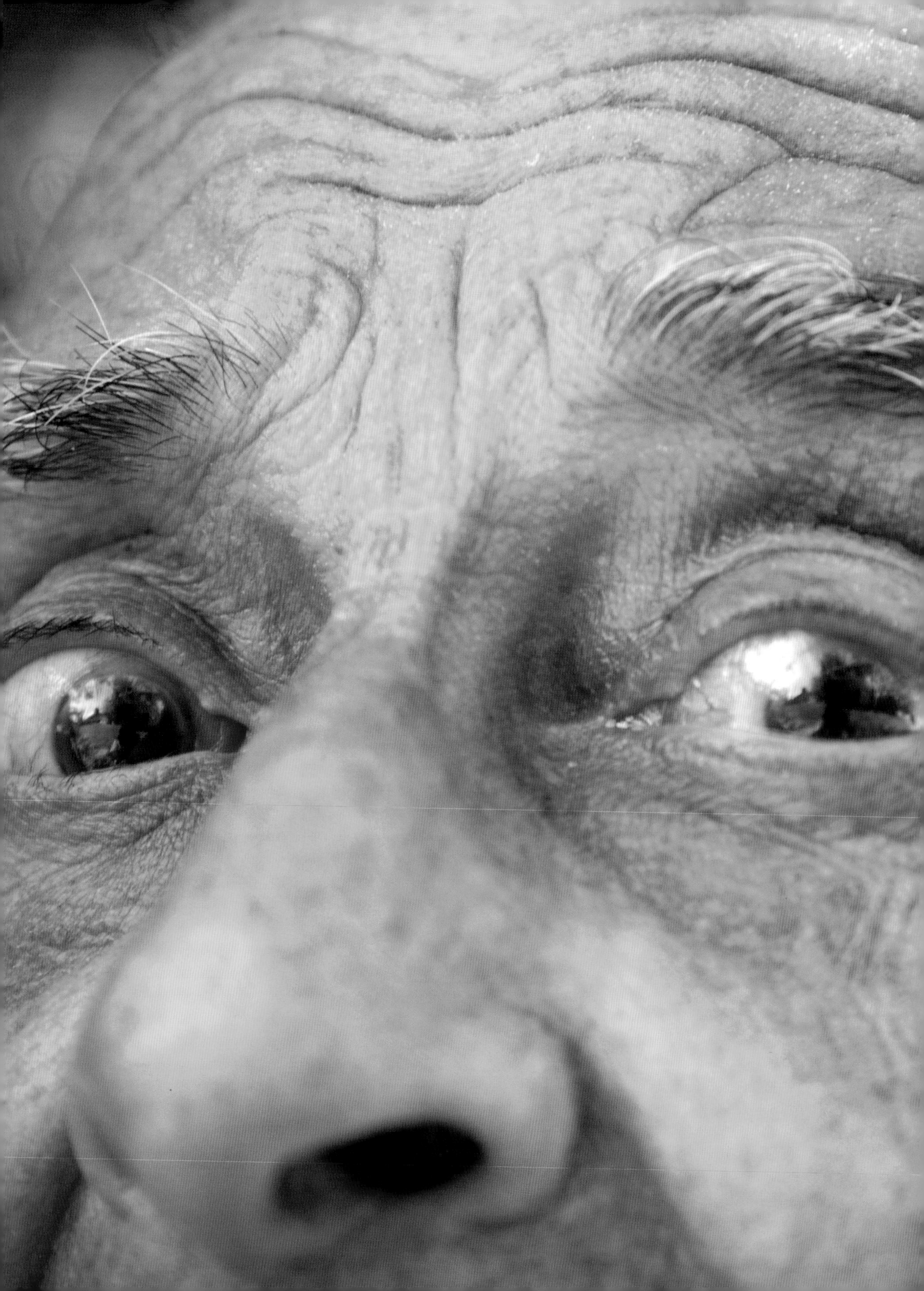

17 CHAPTER *Why be good?*

WHEN *ON THE ORIGIN OF SPECIES* WAS PUBLISHED in 1859, some leading churchmen such as Bishop Samuel Wilberforce and politicians such as Benjamin Disraeli (later to become Prime Minister) immediately deplored Darwin's theory. Their strongest objections were that it placed Man among the animals and replaced morality with naked self-interest. In such a world, why be good? Human society could not function without cooperation among its members, but the existence of unselfish behaviour is an evolutionary puzzle. For centuries believers have worried about why God permits evil. For science the problem is the reverse: how can good behaviour evolve by natural selection?

BELOW Benjamin Disraeli was an instinctive anti-Darwinian. He said in 1864, "The question is this: Is man an ape or an angel? I, my lord, am on the side of the angels".

Good behaviour

One solution to this question was offered by the biologist V.C. Wynne-Edwards, based upon his studies of birds. It is known that in years when resources are scarce, birds of many species lay fewer eggs than they do in years when food is plentiful. Wynne-Edwards argued that this behaviour was bad for individual birds, in the sense that they were not maximizing their individual reproductive rate, but was good for the *population* of birds. The argument goes like this: if there are too many chicks to feed for the resources available, competition is intense, all birds' welfare suffers, and the whole population may go extinct. Populations containing birds that exhibit self-restraint when food is scarce are less likely to go extinct than populations of birds who lay as many eggs as they are individually capable of laying. Thus, by the differential extinction of groups, a kind of goodness evolves, 'goodness' in this context meaning favouring the higher interests of the population above the individual's immediate interest.

The problem with Wynne-Edwards' elegant mechanism for the evolution of goodness is that it doesn't work. Populations containing self-restraining individuals may indeed be less likely to go extinct than ones containing no such individuals. However, *within* those populations, it is individuals practising less self-restraint who leave the most descendants. Thus, over the generations, the proportion of individuals descended from birds who did not practise restraint becomes larger and larger, until all birds are of this type. Behaviour for the higher good tends to be out-competed locally by behaviours that simply maximize individual reproductive success.

OPPOSITE If humans are animals, why be good?

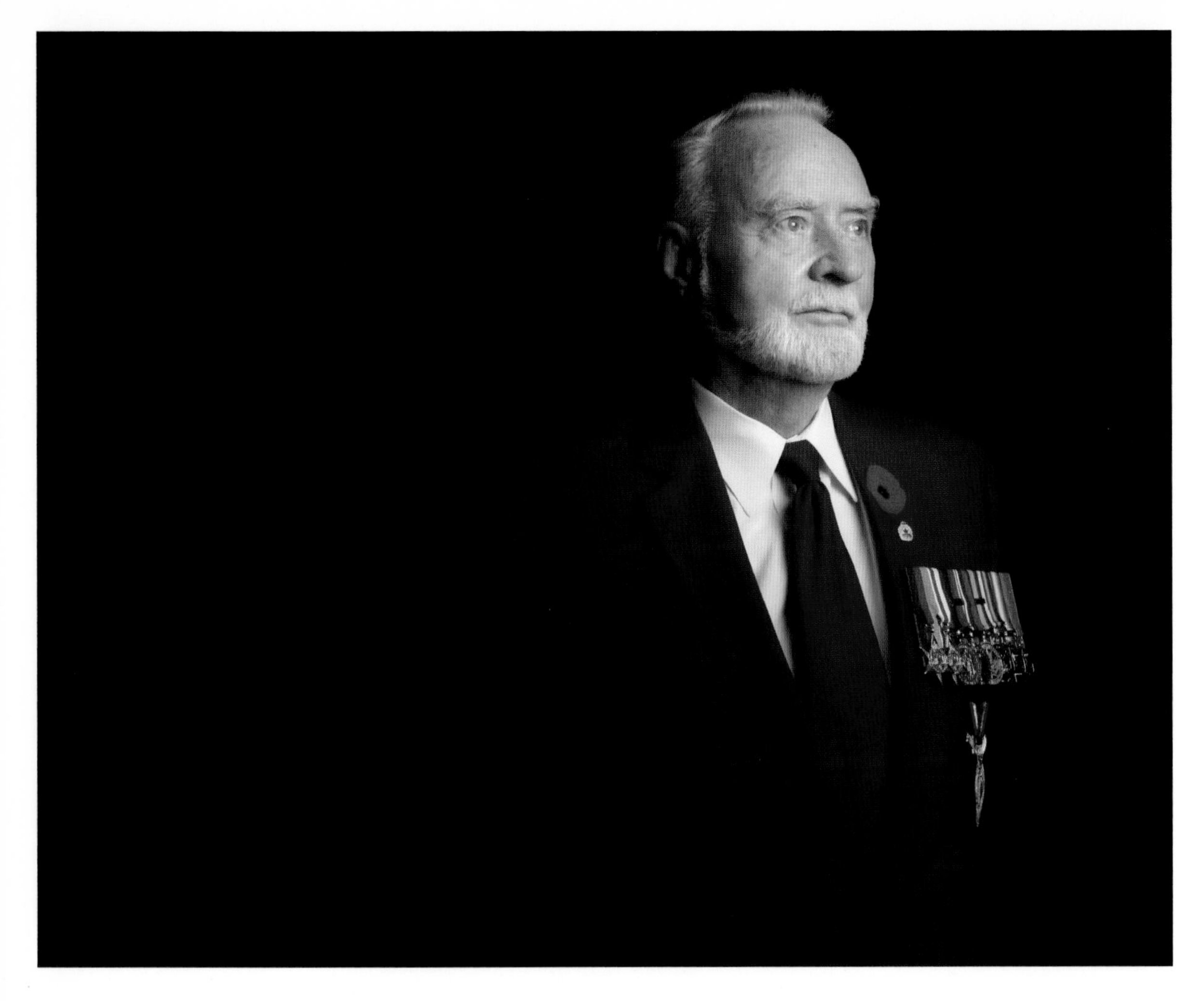

ABOVE Why should people do things for the good of the group but at a cost to themselves. Soldiers risk their lives and receive medals in recognition of their bravery.

The fact that individual interests tend to triumph over the interests of larger groups like populations or species has disquieted some people. The Darwinian world-view dislodges any divine scheme as an explanation of how we behave, and in its place puts the notion that we should perform such behaviours as maximized the reproductive success of our ancestors. If we now concede that the behaviours that triumph in competition are the ones which benefit individuals, not the groups to which those individuals belong, then we *seem* (and the *seem* should be stressed) to have a prescription for a brutish, competitive world where every individual is narrowly focussed on promoting his or her own advantage, or at most that of close kin.

The actual *data* on human behaviour, and indeed many animal societies, do not suggest such a world. Although individuals do sometimes embezzle, steal, murder, and so on, they do not *usually* do so. In fact, antisocial and criminal behaviour, though recurrent through history, seem to be the aberrant case, worthy of special comment and explanation. There is also plenty of evidence of humans routinely doing things to positively benefit non-relatives. Such behaviours are known as pro-social, and there are many examples. People work together on joint ventures, contribute to public services, help strangers, and give to charity.

Gameplay

There is a simple experiment used by behavioural researchers called the *ultimatum game*. There are two players, A and B. A is given a sum of money (say ten dollars) and told to divide it in any way she likes between the two of them. Player B can either accept the proposed division, in which case they both get the allotted sums, or reject it, in which case they both get nothing. The prediction of narrow self-interest is that player A will allot the minimum possible fraction of the money (say, one dollar) to player B, keeping the rest (say, nine dollars) for herself, and B will accept this division, since a small amount is still better than nothing.

ABOVE Adam Smith, best remembered as an economist, pioneered the study of moral intuitions.

The experiment has been performed in many different cultures with varying amounts at stake. Though there is quite a lot of cross-cultural variation, in no culture is the minimal offer the norm. The average share offered is between one quarter and one half, and one of the most common offers is 50%. As for player B, he often refuses low offers, even though as a result he is always worse off than if he accepted.

These results confirm certain principles that philosophers have written about for many centuries. First, it seems normal for human beings to pay attention to the outcomes others receive as a result of their actions, and not just compute their own narrow advantage. Second, people have what Scottish philosopher Adam Smith (1723–1790) called *moral sentiments* about what is good and bad behaviour. Where someone violates a moral sentiment, we wish to shun or punish them. In the ultimatum game, this is evident in the refusals of low offers by player B. Excessively low offers are felt to be bad, and player B is frequently prepared to take a financial loss to signal so to player A.

LEFT If you were given a sum of money to share between yourself and a stranger, how much would you give them? And if you were the stranger, what is the minimum you would accept?

The origin of morals

How have we evolved our pro-social behaviours and moral sentiments? Evolution favours behaviours that benefit the individual, but there are circumstances where the individual benefits from being pro-social. We should expect people to be sensitive to the kind of situation they are in, and to be pro-social where it is in their overall interests to be so. Thus, though evolutionary theory cannot solve the problem of why we ought to be good, a knotty problem that philosophers have been grappling with for centuries, it has a great deal to say about when we are most likely to be good, and why.

The simplest mechanism leading to the evolution of pro-social behaviour is when it coincides with simple self-interest. For example, maintaining hygiene around your home produces a benefit to your whole community, by reducing the spread of diseases. However, it also provides a direct benefit to you, as you are the one most affected by infectious agents nearby, and so it is worth your doing. Biologists call such situations *by-product mutualisms*, because the apparent pro-sociality is really just a by-product of the actor maximizing his advantage. All that is required for a by-product mutualism to work is that the benefit to the actor be greater than the cost. The beneficial effects on others do not affect the decision either way. For example, it could be worth rich individuals paying for public health measures that reduce the level of infectious disease, such as better sewerage, because it benefits them and they have the resources to do it. The fact that many poorer people also benefit, perhaps even benefit more, is a happy side effect of directly self-interested behaviour. However, by-product mutualisms do not explain such situations as the ultimatum game.

ABOVE People sometimes undertake public works, such as here installing London's sewerage system in Victorian times, because of the direct benefits they gain, and sometimes because of the indirect benefits accruing to their reputations.

Another reason why it might be adaptive for individual A to do something to benefit individual B is that individual B might be able to do something beneficial in return, perhaps at a later point in time. Direct reciprocity of this kind can evolve where a number of conditions are met. First, individuals must be able to recognize and remember each other in future, to determine to whom reciprocal behaviours should (or should not) be directed. This restricts direct reciprocity to organisms with considerable cognitive sophistication. Second, both parties must be sometimes in a position to help the other. If one party always has food, and the other never does, then for obvious reasons reciprocity cannot evolve. Finally, the probability of the two individuals meeting again must be high, and the benefit of the act to the recipient must exceed its cost to the donor. Even given these conditions, direct reciprocity is precarious in sizable populations, because of various ways that 'cheaters' (individuals who never return the benefits they receive) can disrupt it.

Direct reciprocity cannot, even in principle, explain all human pro-social behaviour. People give to charity or to programmes that benefit all of society, and contribute to the well being of individuals who are never likely to be in a position to reciprocate, even where there seems to be no immediate personal gain. One possible mechanism maintaining the pro-social behaviour in such instances is *indirect reciprocity*. Indirect reciprocity is where individual A helps B as long as B has helped others in the past. It is not necessary for B to have helped A, or even interacted with A before, as long as A has information about the past behaviour of B. Indirect reciprocity is, in other words, a system of reputation. By their good actions, individuals can cultivate a reputation that means they receive social benefits that exceed the costs of their pro-sociality. Conversely, if they cheat on others, they form a bad reputation that means they are excluded from social benefits to an extent that outweighs the short-term benefits of cheating.

Indirect reciprocity also has conditions for stability. It can spread if the probability that individuals have accurate information about each other's prior behaviour is high. Gathering accurate information about the prior behaviour of everyone in one's social environment at first hand would take a lot of time and effort, which seems to place a limit on the evolution of indirect reciprocity. However, humans have a special adaptation which helps solve this problem: language. Language allows for the low-cost transmission of information about the behaviours of third parties, and as a result can make indirect reciprocity much more robust. Observational studies of language use confirm that gossip – essentially information about the past behaviour of others in the social network – is the single most frequent category of information in conversations in many contexts, and more cooperation is found in experimental cooperation games where the players are allowed to talk to each other about third parties than where they are not.

The hypothesis that indirect reciprocity plays a key role in human pro-social behaviour leads to a number of predictions. First, people who are visibly pro-social should receive more benefits from third parties than those who are not. Second, seeing (or hearing via language) that an individual has been anti-social with a third party should make us reluctant to do anything to their benefit. Third, people should be more pro-social where others are likely to come to know about their actions than where they do not.

Researchers have been able to test all of these predictions experimentally and found support for all of them. In one experiment, participants played a game where they could choose whether or not to confer a monetary benefit on another player (at cost to themselves). Participants were anonymous, and were explicitly told that the other player would not be in a position to reciprocate. However, they would interact with further individuals, and those individuals would know exactly what they had done in earlier interactions. Not only was there a high frequency of giving, but individuals who were more generous received more benefits later on, from third parties who had observed their generosity, and individuals who were not generous early on reaped the consequences.

In the public eye

These findings have been extended to 'public goods' situations. A public good is something that benefits all individuals, regardless of whether they have contributed to its costs or not. Probably the most obvious public good affecting humanity today is the atmosphere. If some people emit fewer polluting gases into it, then those who have cut their emissions and those who have not benefit alike. The person who profits most in this situation is the person who does not cut his own emissions while everyone else does, since he gets all of the benefits and none of the costs. Because of this perverse incentive, societies tend to provide lower than optimum levels of public goods, unless there is some legal mechanism enforcing equal contributions.

In another experiment, participants were allowed to invest money in a 'climate change' fund, which would be invested in reducing fossil fuel use, or to keep it for themselves. They then interacted individually as before. In one condition, whether they had contributed to the climate change fund was secret, and in another, their future interaction partners knew the information. Where contributions to the climate change fund were public, they were always higher than where they were anonymous, and moreover, individuals who did not publicly contribute to the fund were punished by receiving fewer benefits in the individual interactions that followed than those who did.

These are very exciting findings. They suggest that as long as there is public visibility of actions, then it can be in individuals' self-interest to be good because of the reputational benefits being good brings. This means that good actions may arise spontaneously and without the need for costly and difficult legal enforcement, as long as there are reputational consequences of behaviour. This principle has been discovered by online commerce sites such as eBay. Given the vast numbers of buyers and sellers on eBay, their lack of real names, and their geographical spread, it is very difficult to police the transactions in the conventional way. Instead, eBay uses a reputation scoring system where any buyer can make public the level of service she has received. Sellers with good reputations actually receive higher prices for their goods, and thus it becomes in their direct self-interest to be honest.

Home > Community > Feedback Forum > Feedback Profile

Feedback Profile

cckjtech (351 ★)

Member since 25-Jul-07 in China

Contact mem

Feedback Score: 351
Positive Feedback: 100%

Members who left a positive: 351
Members who left a negative: 0

All positive Feedback: 415

Find out what these numbers mean

Recent Feedback Ratings (last 12 months)

	1 month	6 months	12 months
Positive	303	415	415
Neutral	1	1	1
Negative	0	0	0

RIGHT Online trading sites such as eBay exploit the power of reputation in maintaining and enforcing cooperative behaviour, by allowing any buyer to make public the level of service s/he has received.

LEFT The feeling of being watched makes people concerned for their reputations.

To make people more pro-social, it may not be necessary to actually increase the reputational consequences of their behaviour. It often suffices to make them *feel* like their behaviour is highly visible. For example, over evolutionary time periods, the proximity of human eyes pointed in our direction (a signal enhanced by the whites of our eyes – see box p.191) has generally meant that people are watching, people who judge and talk about what we do. Generosity in experimental cooperation games can be increased merely by including eye-like stimuli on the screen background of the computer at which the person is completing a task. The cue uses the powerful role that eyes play in social interactions and subtly evokes the psychology of reputation, even though no actual watching is taking place.

This experiment was recently replicated in the coffee area of a university department. Coffee was paid for by placing cash in an honesty box, and the prices and instructions were displayed on a piece of paper stuck to a cupboard door. In some weeks, a picture of human eyes was added to these instructions, with a picture of flowers in the intervening ones. In weeks with eyes, substantially increased contributions were received for the amount of coffee consumed.

In conclusion, what does evolutionary theory say about the age-old question of whether humans are basically good or basically bad? It says that they are not inflexibly one or the other. Rather, under some circumstances their best interests are served by being pro-social, and they should have evolved to respond to these circumstances with appropriate pro-social behaviour. Empirical evidence suggests that they do. Circumstances favouring pro-sociality include where there is a direct benefit from being part of a functional social network, or where there are likely to be reciprocal or reputational payoffs for being good. If we wish to maximize pro-sociality, we need to set society up in such a way as to maximize the frequency with which people find themselves in such circumstances. Informal systems of reputation seem to work well to maintain pro-social behaviour in small communities, and presumably sufficed to maintain the social order

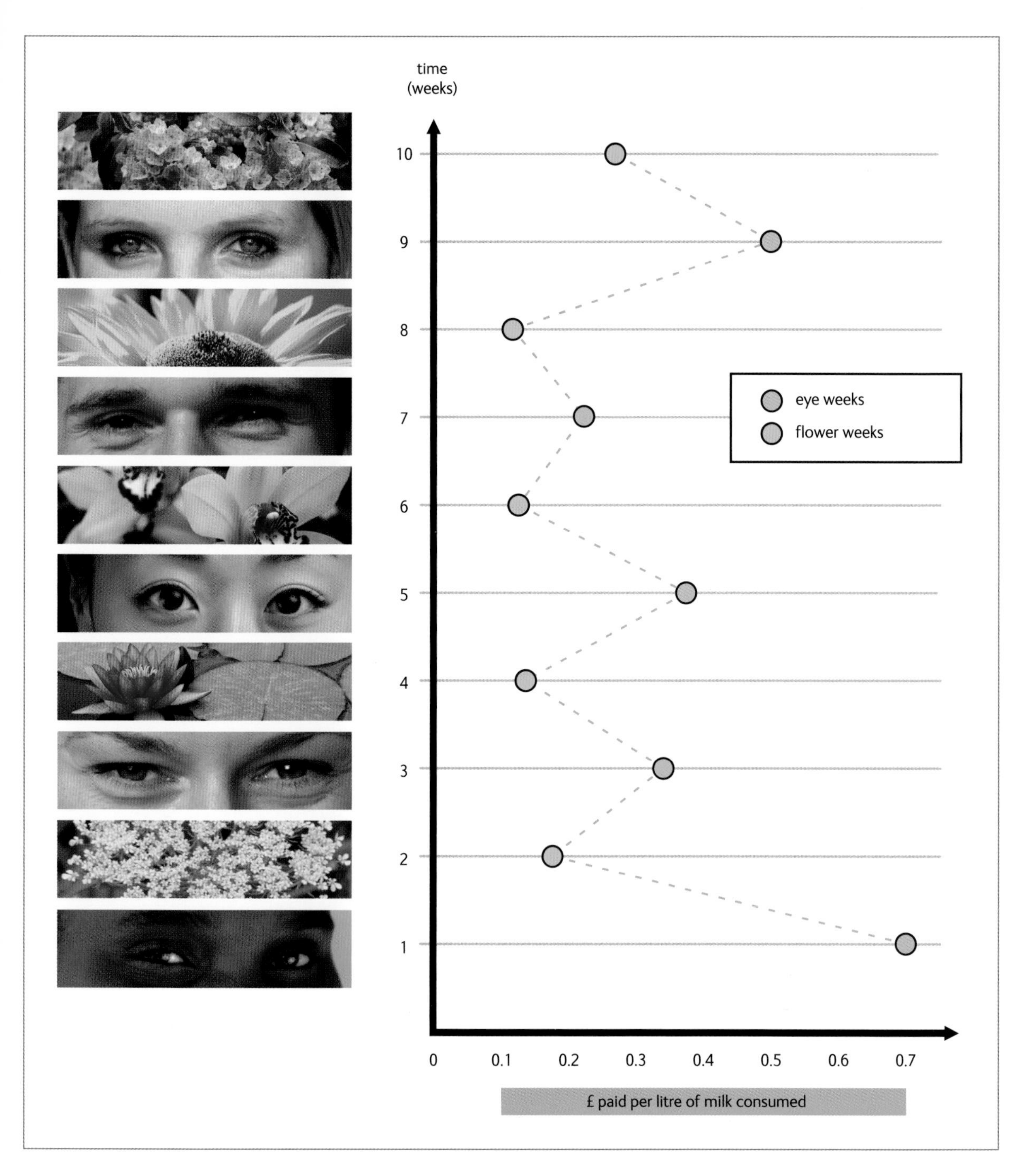

ABOVE Contributions to an honesty box in a coffee room were higher in the weeks when there was a picture of eyes on the wall, than in weeks when there was a picture of flowers.

in the hunter–gatherer societies of our ancestors. For our modern states with millions of citizens, they are not by themselves sufficient – formal policing and punishment are required too – but they certainly help.

Evolutionary psychology is at the frontier of the modern science of evolution. It applies Darwinian thinking to subjects that some would argue are outside its proper province. This raises the question, what is the proper domain of science? Which questions can it answer, and which can it not?

FOR AS YOU WERE WHEN FIRST YOUR EYE I EYED

To me, fair friend, you never can be old,
For as you were when first your eye I eyed,
Such seems your beauty still.

Shakespeare, W. (c. 1600) Sonnet 104

Why are the eyes so expressive of emotion and attention? Darwin discussed this issue at length in his book about the expression of emotions in people and animals. We owe our ability to see well in a range of light levels to almost instantaneous, subconscious adaptations to light, including dilation or constriction of the iris, which admits more or less light onto the retina. Other neural signals and certain drugs have similar effects. The sun seems to shine more brightly when we are in love because the emotion dilates the iris more than usual. As well as experiencing it ourselves, we recognize iris dilation in other people and interpret it as a sign of interest.

In many social animals including dogs, the tissues around the eye also signal mood; half-closed eyelids mean anger or threat. Such signals are enhanced in humans and many other primates by eyebrows, conspicuous bands of hair on bony ridges above the eyes moved by muscles in the skin, whose principal, probably sole, function is social communication. They are absent in eyelidless snakes, whose immobile stare seems to us alien and frightening. Eye contact and eye signals such as winking are learnt early, usually from other family members, and often persist throughout life, creating characteristic poise and mannerisms.

The 'whites of the eyes' around the iris (un-pigmented areas of the fibrous outer layer of the eyeball) facilitate recognition from a distance of people's eye movements and their direction of gaze. We are remarkably good at noticing other people noticing us and reporting what they are looking at. Most people find squints or defects and wounds on the whites of the eyes at least transiently disturbing. Most mammals, including other apes, do not have conspicuous white of eyes; the backs of the eyeball are white but all of the exposed portion is pigmented. However, a few chimpanzees have intriguingly human-looking whites to their eyes, suggesting that a similar heritable change must have been strongly selected in our own ancestors. Some biologists think that this distinctively human feature evolved as part of our elaborate social and sexual behaviour. Similar adaptations have evolved in several groups of social birds, including parrots; they can follow each other's gaze and, with experience, that of humans.

BELOW Chimpanzee eyes are normally pigmented, in contrast to those of humans.

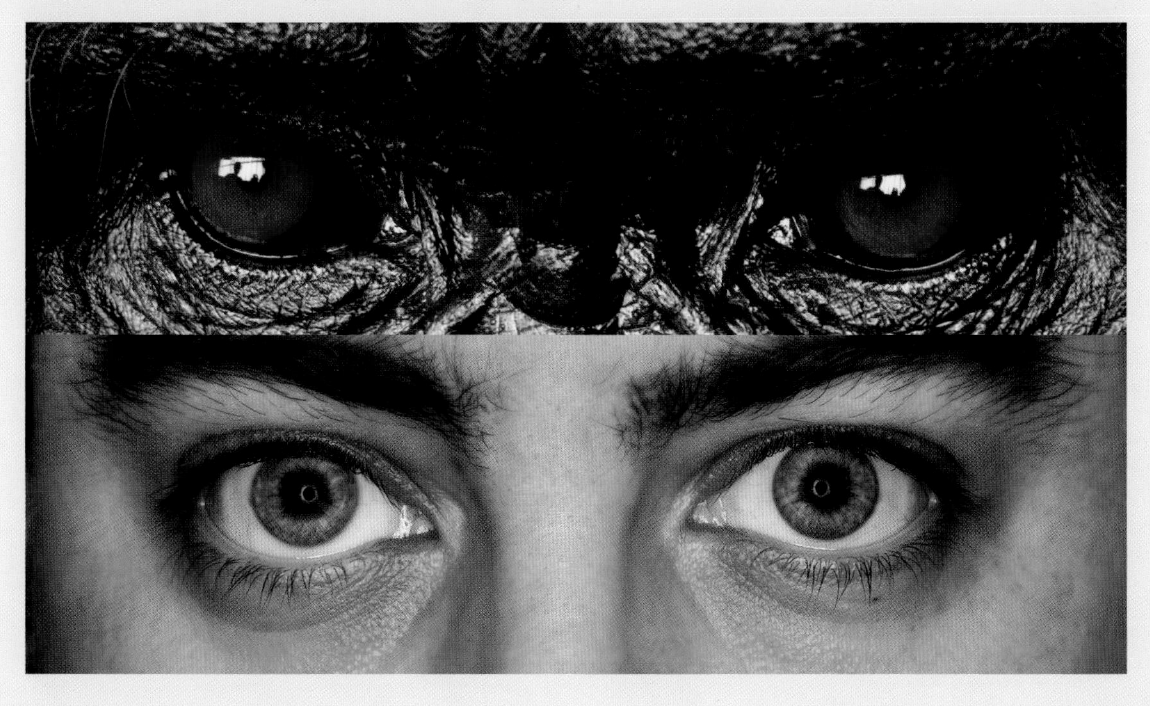

Index Germination of Seeds

Lythrum 1. — 1
Red equal-styled Cowslip A 1
Antirrhinum peloric 2.
Myosotis 2. — 2
Cowslip Pol. 3. — 3
Mimulus 5. — 5
Calceolaria — 6
Canna — 7
Pulmonaria — 8
Red equal styled Cowslip 9
Ipomea — 10
Verbascum thapsi — 11
Primula elatior — 13
Pea common 14
Candytuft white — 15
Candytuft crimson — 16
Sweet Pea — 17
Primrose — 19
Cabbage — 20
Lettuce — 22
Verbascum lychnitis — 23
Petunia — 25
Lobelia ramosa — 27

Marjoram — 2
Cucumber — 29
Lupinus luteus — 3.
pilosus — 32
Peas Laxton's Supd — 3.
Parsley — 35
Carnation — 37.
Cyclamen persica 39
Anagallis grandiflora 41
Bartonia aurea 43
Nemophila insignis 45.
Salvia coccinea 47.
Eschscholtzia 49
Pelargonium — 51
Passiflora gracilis 53.
Primula sinensis 54.
Lobelia fulgens 56
Viola tricolor — 58
Cineraria 60
Thunbergia — 62

18 CHAPTER *The* science *of evolution*

OPPOSITE Charles Darwin conducted many scientific experiments and recorded the results meticulously in his notebooks.

> *'Modern scientific discoveries reveal over and over again that the popular belief associating Darwinism with science is false. Scientific evidence refutes Darwinism comprehensively and reveals that the origin of our existence is not evolution but creation. God has created the universe, all living things and man.'*
>
> Yahya, H. (2006) *Atlas of Creation*, Volume 1

Comments like this are commonplace in anti-evolutionary publications, though this particular example is interesting because it comes from an Islamic creationist, demonstrating that such misinterpretation of scientific evidence is not limited to fundamentalist Christians. So it is worth following the exploration of various evolutionary case histories in previous chapters with a second look at the scientific credentials of the theory.

Evolution versus creation: science and non-science

Science aims to extend our understanding of natural phenomena through testing of explanatory hypotheses by reference to hard evidence. It is not concerned with ideas that cannot be tested in this way, such as subjective opinions (for example, what is good or evil, beautiful or ugly) or religious beliefs (about, say, 'the meaning of life' and the existence of gods or spirits), though we will return to ideas like this at the end of this chapter. The remit of science was eloquently summarized by Judge John Jones III in his judgement on a case heard in Dover, Pennsylvania, in 2005, involving the teaching of evolution in schools (Kitzmiller v. Dover):

> *'Expert testimony reveals that since the scientific revolution of the 16th and 17th centuries, science has been limited to the search for natural causes to explain natural phenomena. This revolution entailed the rejection of the appeal to authority, and by extension, revelation, in favor of empirical evidence. Since that time period, science has been a discipline in which testability, rather than any ecclesiastical authority or philosophical coherence, has been the measure of a scientific idea's worth. In deliberately omitting theological or 'ultimate' explanations for the existence or characteristics of the natural world, science does not consider*

issues of 'meaning' and 'purpose' in the world. While supernatural explanations may be important and have merit, they are not part of science. This self-imposed convention of science, which limits inquiry to testable, natural explanations about the natural world ... requires scientists to seek explanations in the world around us based upon what we can observe, test, replicate, and verify.'

To be worthwhile, moreover, a scientific theory must do more than just 'be testable': it should successfully explain a wide range of phenomena that would be unintelligible or inconsistent under alternative hypotheses. On these grounds, the modern theory of 'variational' evolution, which combines Darwin's theory of evolution by means of natural selection with the discoveries of genetics, is resoundingly successful. Not only is it entirely naturalistic, such that every component can be (and has been) tested, but it also explains a remarkable variety of biological phenomena, including some that would otherwise seem positively inexplicable. We have seen in previous chapters how it elucidates the diversity and distribution of organisms in space and time, the origins and nature of adaptations (their appearance of intricate design, but common lack of perfection), the existence of shared inherited characters, convergence and vestigial organs, the origins of species, including our own, the natural history of diseases, and even the way our minds work. As one of the leading evolutionary biologists of the last century, Theodosius Dobzhansky (1900–1975), remarked:

'Nothing in biology makes sense except in the light of evolution'.

American Biology Teacher (1973)

BELOW Special creation, a vacuous and unscientific hypothesis, no matter how beautiful.

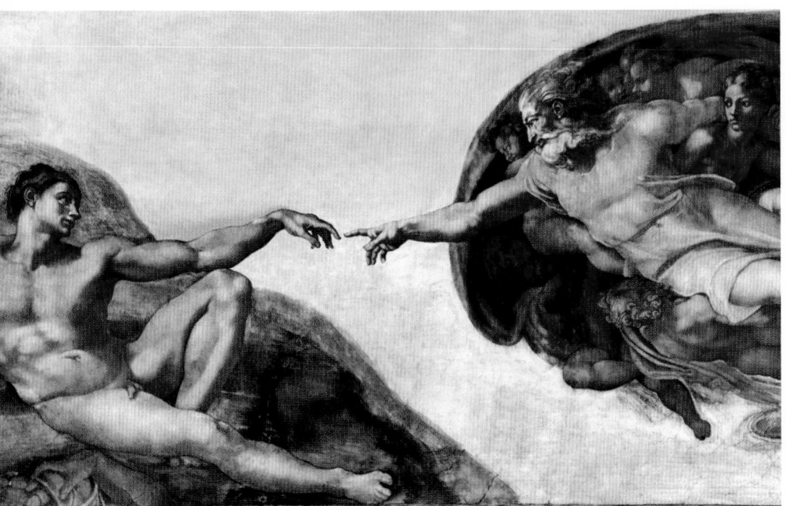

By contrast, the hypothesis of special creation not only flies in the face of the evidence, in so far as it can be tested (Chapter 3 *The tree of life*), but in consigning the creative process to the unknowable the hypothesis doesn't explain anything beyond the mere existence of species. So it is not only a false hypothesis, but a vacuous one. And finally, its unverifiable supernatural element (the purported 'creator') puts it beyond the purview of science in any case.

This distinction between evolutionary science and creationist non-science is of more than just academic importance. Predictions concerning the impact on life of rapid climate change and other perturbations of global environments, as well as proposals for mitigation, crucially depend on a sound theoretical framework. For the reasons given above, special creation cannot be accepted as an alternative scientific theory on an equal footing to modern evolutionary theory, and so it has no legitimate place in any science classroom. It is in this context that the question of what qualifies as science, or not, has on a number of occasions come under legal scrutiny, especially in the United States, where a vociferous lobby of fundamentalist Christian creationists has repeatedly attempted to force their doctrine onto school science curricula at the expense of sound biology teaching. Fortunately, so far, their efforts have been checked by a number of astute legal judgements and unmasked for what they really are – covert attempts to introduce religious teaching into science classes, where it not only has no legitimate place, but would actually contravene US Constitutional law. Unfortunately, similar legal safeguards against the subversion of science education seem to be lacking in Britain.

Creationism in disguise

In recent years, creationists have re-branded their hypothesis as 'intelligent design', which asserts that the apparently designed fit of organisms to their conditions of life necessarily implies the existence of an intelligent designer. This idea is no more than the old 'argument from design', promulgated most famously by William Paley in his fable of chancing upon a watch and inferring a watchmaker (Chapter 2 *Darwin's brilliant idea*), which the theory of evolution by natural selection refutes, as brilliantly explained by Richard Dawkins in his aptly titled book *The Blind Watchmaker*. Paley's analogy is false, because watches are not living organisms capable of reproduction, hence evolution. Needless to say, 'intelligent design' has not fooled the American courts, who have recognized it simply as creationism in disguise.

Evolution and religion

It would be a mistake, however, to suppose that the invalidity of the creationist hypothesis and the redundancy of any supernatural agency in the Darwinian theory of evolution by natural selection necessarily endorse atheism, as has been mistakenly supposed both by creationists and some evolutionists alike. That God has not miraculously created every species does not disprove the existence of a deity. In fact, Darwinian evolutionary theory is equally compatible with theism and atheism. Darwin himself eventually became an agnostic, correctly recognizing such a personal issue as being irrelevant to his science. In this context, it is of interest to note the following pronouncements from Pope John Paul II in the late 1990s:

> *'It may help, then, to turn briefly to the different modes of truth. Most of them depend upon immediate evidence or are confirmed by experimentation. This is the mode of truth proper to everyday life and to scientific research. At another level we find philosophical truth, attained by means of the speculative powers of the human intellect. Finally, there are religious truths which are to some degree grounded in philosophy, and which we find in the answers which the different religious traditions offer to the ultimate questions.*

> *… Even if faith is superior to reason there can never be a true divergence between faith and reason, since the same God who reveals the mysteries and bestows the gift of faith has also placed in the human spirit the light of reason. This God could not deny himself, nor could the truth ever contradict the truth.*’
>
> Pope John Paul II (1998)

Leaving aside the philosophical problem of what religious ‘truth’ means, the key point of interest here is the assertion that the articles of faith and the well-established findings of science cannot flatly contradict one another. Hence, the findings of science concerning evolution received explicit papal endorsement:

> ‘*In his encyclical Humani Generis (1950), my predecessor Pius XII has already affirmed that there is no conflict between evolution and the doctrine of the faith regarding man and his vocation, provided that we do not lose sight of certain fixed points …*
>
> *Today, more than a half-century after the appearance of that encyclical, some new findings lead us toward the recognition of evolution as more than a hypothesis. In fact, it is remarkable that this theory has had progressively greater influence on the spirit of researchers, following a series of discoveries in different scholarly disciplines. The convergence in the results of these independent studies – which was neither planned nor sought – constitutes in itself a significant argument in favour of the theory.*’
>
> Pope John Paul II (1996)

Logos and mythos

An interesting perspective on the distinction between religious belief and scientific understanding has been provided by Karen Armstrong, who relates them to two complementary modes of human thinking, both with ancient pedigrees and given the classical Greek names of ‘*logos*’ and ‘*mythos*’. The former deals with the practical understanding of how nature works, and has long been used to advantage in, say, agriculture and technology. Although this mode of thinking, as exemplified by modern science, can satisfy our natural curiosity concerning objective matters, it cannot, as noted earlier, fully address our subjective concerns with ethical values, aesthetic judgements and any personal sense of identity and purpose in life, although it may inform our views on them.

Such irrational needs have, throughout history, been ministered to by the various forms of *mythos*. The point of *mythos*, Armstrong argues, is not literal explanation, which is what *logos* provides, but – through symbols, sagas and rituals – to inspire a sense of seeing beyond mundane matters, so to invest life with meaning and value. Hence to expect *mythos* to furnish informative answers to questions that are properly the domain of *logos*, such as the origins of life’s diversity and adaptedness, and indeed of ourselves, is to confuse the psychological roles of the two modes of thought. Yet that is precisely the confusion to which creationists of various 'fundamentalist' denominations have succumbed, as an essentially modern – one might say paranoid – reaction to the ascendancy of science and retreat of religion over the last few centuries.

The irony of the creationists' self-delusion is that, with a little more study of the very scriptural sources to which they appeal, they would soon discover something that has long been known to biblical scholars: inconsistencies in the texts themselves render insistence on literal interpretation logically untenable, without the need even for scientific refutation. Thus, for example, besides the two incompatible creation myths that follow one another in the Book of Genesis, there are further allusions in subsequent books to yet another contrasting myth involving an initial struggle between Yahweh and a primal sea monster, inherited from Babylonian mythology.

ABOVE The start of only the first of the biblical creation myths.

As an apt response to the quotation at the start of this chapter, then, we can leave the last words to Karen Armstrong:

> *Creation stories have never been regarded as historically accurate; their purpose was therapeutic. But once you start reading Genesis as scientifically valid, you have bad science and bad religion.*
>
> Armstrong, K. (2005) *A Short History of Myth*

Creationism is not really a significant challenge to evolution because, as we have seen throughout this book, the scientific evidence for evolution is overwhelming. A more interesting challenge for evolution as a science is whether it can tell us anything about the future.

THE SCOPES MONKEY TRIAL

In 2007, Professor Michael Reiss, a Church of England priest and the head of science at London's Institute of Education, said that it is becoming more difficult to teach evolution in schools because of the spread of creationism. Similar debate has long been burning in the United States. Also in 2007, a creationist museum opened near Cincinnati, where children in animal skins play amid model dinosaurs, suggesting they once coexisted and that the geological timescale is nonsense. The museum's aim is to bring Genesis – the first book of the Bible – to life for all ages, and promote the belief that the Earth is less than 10,000 years old.

When a classroom debate becomes a courtroom debate, the law has a solemn responsibility. Schools, after all, are engaged in the mass production of the minds of the future. Evolutionary science was at the heart of one of the most famous cases involving a school curriculum: the 'Scopes monkey trial', which took place in Dayton, Tennessee, in 1925 and featured one of the United States' most famous lawyers, Clarence Darrow.

The case arose from events stirred by the passing of a law in Tennessee making it illegal for any publicly funded institution to deny the Bible's theory of creation and to teach instead that humans descended from 'a lower order of animals'. The law was promoted by William Jennings Bryan, a lawyer, fundamentalist Christian and former presidential candidate who wanted to banish Darwinism from classrooms. The film *Inherit the Wind* is loosely based on the real case but makes several significant departures from the real events.

John Scopes, who had taught some science classes on a part-time basis, agreed to teach some biology classes in order to set up a test case. So, as part of those classes he taught the theory of evolution. William Jennings Bryan, having championed the legislation, volunteered to prosecute him. Clarence Darrow, an agnostic, liberal, and a famed defence lawyer, represented him.

The trial took place in a sort of carnival atmosphere in July 1925. There were banners in the streets, lemonade stalls, chimpanzees performing in a sideshow. A thousand people, 300 of them standing, packed into the Rhea County Courthouse. Soon after it began, the trial was moved outdoors and the crowd grew to 5000. It was the first trial attended by radio announcers: they gave live updating broadcasts to listeners. Presiding at the trial was Judge John T. Raulston, a conservative Christian. As often happened, the proceedings were opened with a prayer.

The prosecution's case was put simply, by showing that Scopes taught the forbidden science. The trial became lively

ABOVE Creationists believe humans and dinosaurs existed at the same time.

when Darrow began the defence. His first witness was Dr Maynard Metcalf, a zoologist from Johns Hopkins University, who explained the Darwinian theory of evolution. Bryan, following Dr Metcalf's testimony, expressed his disgust at the notion that man evolved 'not even from American monkeys, but Old World monkeys'.

In an extraordinary scene, the defence called the prosecutor himself to give evidence on the witness stand as a bible expert. As a witness, Bryan was asked questions by Darrow about Jonah being swallowed by a whale, Joshua making the Sun stand still (how so when the Earth moves round the Sun?), and the Earth being created in one week. Bryan said:

> *"Everything in the Bible should be accepted as it is given there."*

He later gave way on some literalist interpretations and stumbled in some of his answers:

> *"I do not think about things I don't think about."*

If the great flood that destroyed all civilizations was in 2348 BC, how did Bryan explain civilizations that traced their history over 5000 years? One exchange went like this:

> DARROW: *"Have you ever investigated to find out how long man has been on the Earth?"*
>
> BRYAN: *"I have never found it necessary."*

Eventually, the judge, who had been in church in the front pew when prosecutor Bryan had delivered a sermon on the first Sunday during the trial, stopped Bryan's evidence and adjourned the hearing. He then ordered Bryan not to return to the stand, and ordered his earlier testimony to be stricken from the record.

At the end of the trial, Darrow asked the jury to return a verdict of guilty so that the case could go to the Tennessee Supreme Court. The jury obliged, and Scopes was fined $100. The conviction, though, was overturned on appeal – not on the constitutional basis on which Darrow and his team had argued (freedom of expression for science) but because the judge had fixed the $100 fine. According to Tennessee law, since it was over $50, it should have been fixed by the jury. So the defendant was acquitted on this technical matter about the fine, not because it was judged as proper for science to be a better way of teaching biology than creationism. No other prosecutions were ever brought under the Tennessee law, and it was abolished in 1967.

ABOVE Clarence Darrow and William Jennings Bryan at the trial.

The law court, both here and in the Kitzmiller v. Dover (2005) case (see p.193), was a good forum for the debate about whether religious creationism should be taught as part of science. Law courts are good forums for debates because they have sensible rules about people using witnesses, and the witnesses for each side being open to be cross-questioned by the other side. A legal setting provides a great structure in which to conduct a rational argument in fair conditions. Not everyone though is sympathetic to lawyers. Speaking about the Scopes trial, the American humorist Will Rogers said:

> *"We had that monkey trial down in Tennessee to prove that man descended from the apes but I never believed that. Because I never yet met an ape who was devious, heartless or greedy, I always figured man was descended from lawyers!"*

19 CHAPTER What next?

WHEN THE BERLIN WALL FELL IN 1989, AND the Soviet Bloc began to disintegrate and the Cold War between East and West ended, the American political theorist Francis Fukuyama penned an essay entitled 'The end of history?' The end was marked by:

> '... *the end point of mankind's ideological evolution and the universalization of Western liberal democracy as the final form of human government.*'

Subsequent events have, of course, demonstrated that competing ideologies continue to thrive. It now seems obvious that history will only cease with the end of human existence. The idea that history had ended was itself the product of a particular, triumphalist moment that has now passed. History may not have ended with the victory of the West over the East in the Cold War, but maybe victory in another war has brought human evolution to a stop. Does technology now shield us from the forces of heritable change and natural selection?

OPPOSITE Social networks engender rapid culural change.

Are we still evolving?

Now that we are able to control our own health and fertility, has the struggle between humans and nature been all but won? When we can conceive children with partners that we have never even met or, as seems possible in the near future, we can clone our cells to repair our bodies, are we so in control of our own destiny that there is no room left for natural selection? Can we still be evolving, or is our genome now suspended, as though weightless, in a selection-free environment cushioned by medicines and public health measures that have cut infant mortality rates and lengthened adult life expectancy? Perhaps our capacity to read the human genome sequence means that we can now freeze the text by editing out harmful mutations or render those mutations harmless with personalized medication. Has cultural evolution now replaced biological evolution? When social developments, such as worldwide networking over the internet, can change our habits and opportunities in a few years, can biological evolution operating at the measured pace of generations possibly be significant any longer?

Modern biology and medicine are genuine triumphs of science that have the capacity to help end starvation and to eradicate mass killers like malaria, AIDS and tuberculosis. Smallpox has already been globally eradicated. Is it more than a matter of time before other scourges are also removed, even from the poorest people on Earth? The science fiction writer William Gibson is often quoted as saying:

> '*The future is already here – it is just unevenly distributed.*'

What he meant by this is that, for example, the standards of health that are currently enjoyed by people in Japan, say, where life expectancy is higher than anywhere else, will one day be universal.

On the other hand, perhaps we should learn from Francis Fukuyama not to be misled in a moment of triumphalism. Starvation and disease have been major agents of natural selection in the recent past (Chapter 15 *Catch me if you can*) and have not been defeated, even in rich nations. Emerging infectious diseases introduce new sources of natural selection. Technology may colour or alter the direction of natural selection, like light passing through a prism, but evolution will continue along one path or another. In fact, natural selection may even be accelerating.

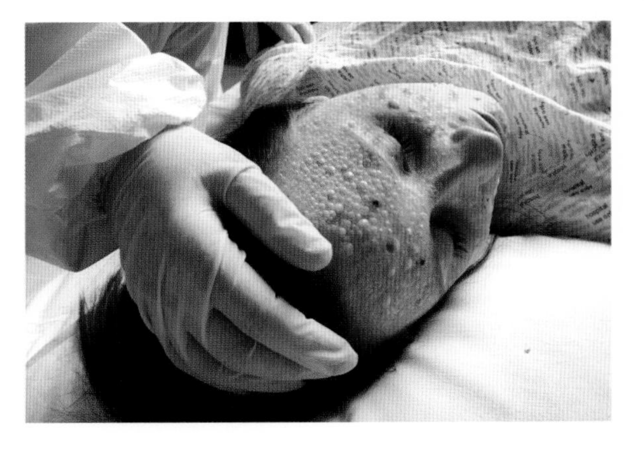

Sheer weight of numbers dictates that favourable mutations, from which natural selection can mould new adaptations, are more likely to appear in a large population than a small one. For this reason, human evolution was accelerated over the last 10,000 years by the combination of a rapid increase in the human population and the many different environments in which we lived, offering new opportunities for natural selection to produce locally favourable adaptation. For example, the advent of cereal farming in the Neolithic increased the supply of energy-rich carbohydrates available to people who had previously been hunter–gatherers. A consequence was that natural selection increased the number of copies of genes coding for starch-digesting enzymes in the farming population.

Many polymorphisms present in our species, like alleles affecting skin colour (Chapter 10 *In the genes*), are recently evolved. Since the human population continues to increase and may reach 9 or even 10 billion before it eventually stabilizes, there is no reason to think that humanity has taken its foot off the accelerator pedal of evolution. Increasing population density itself produces new challenges to health and wellbeing, both directly and indirectly through the effect of an increasing population upon our environment.

OPPOSITE Two generations of Japanese people.

ABOVE Simulation of a patient infected with smallpox.

BELOW Cereal agriculture changed the diet of farming populations and led to evolution in the human genes required for digesting starch.

Old age – when natural selection retires

Natural selection gradually weeds out from a population congenital defects that lower reproductive success, but there are situations when this appears not to happen. More common congenital conditions like sickle cell anaemia are often side effects of beneficial mutations that have been favoured by natural selection in the past. Sickle cell anaemia, thalassemia and some other conditions are the side effects of alleles that have been favoured by natural selection because they protect against malaria. This explains why sickle cell anaemia and thalassemia are most common in people of African or Mediterranean descent whose ancestors were frequently exposed to the malarial pathogens. Certain other genetic diseases, like cystic fibrosis and haemophilia, result from recurring mutations that natural selection is slow to purge because the genes can be transmitted by symptomless carriers. Everyone is prey to another group of inherited conditions that natural selection is unable to purge: the degenerative disease, dementia and decrepitude that may come with old age.

Like the dusty corner of a room that no broom ever reaches, old age is a repository for deleterious conditions that natural selection cannot touch. Natural selection is ineffective against diseases that express themselves only later in life because these conditions do not impair reproductive success. Older individuals have already produced

BELOW AND OPPOSITE Charles Darwin at ages 7, 35 and 70.

SPECTACLES AND EVOLUTION

John Maynard Smith (1920–2004), one of the greatest evolutionary scientists of the 20th century, had extraordinary insight but dreadful eyesight and he used this combination to effect. He explained to students that whereas in human pre-history extreme short-sightedness would probably have led to an early death, his own myopia had saved him from conscription into the army in World War II and therefore possibly saved his life. The lesson was that a trait that is disadvantageous in some circumstances, can become advantageous in others. Changing circumstances, particularly when a trait is affected by technology (for example, spectacles), make it almost impossible to predict the future direction of evolution.

In the 1920s and 30s, there was a small but influential movement of intellectuals in Britain who advocated the application of selective breeding to the human population to improve the 'genetic stock', in the same way as artificial selection improves domesticated livestock. The policy was called eugenics. One of its strongest supporters, Marie Stopes (1880–1958), actually severed all contact with her own son when he contravened her eugenic principles by marrying a woman who wore glasses. Support for eugenics in Britain ended when it became associated with the Nazis' racial policies in Germany. Ironically, the glasses-wearing woman of whom Marie Stopes so disapproved was the daughter of Barnes Wallis, the British engineer who invented dam-busting bombs that helped defeat Nazi Germany.

The last word on the subject goes to John Hegley, a poet whose work celebrates the mundane and ordinary, and – in particular – spectacles. To him, wearing glasses is:

> '*... a symbolic celebration of the wider imperfection that is the human condition*'

TOP John Maynard Smith (1920-2004), eminent evolutionary biologist.

ABOVE Marie Stopes (1880-1958).

RIGHT John Hegley, poet and spectacles-wearer.

all or most of their offspring and have therefore transmitted their genes – including any genes that may lead to malfunction in old age. Therefore, mutations that express themselves only late in life tend to accumulate over evolutionary time, leading to the evolution of ageing. For the same reason, mutations that increase reproductive success early in life, but also have deleterious consequences for the carrier of those genes in later life, are favoured. For example, the inflammatory response with which our immune system protects us from infection and which is vital to survival in early life may also play a part in causing the degenerative diseases of old age such as arthrosclerosis, osteoarthritis, osteoporosis and Type II diabetes.

Degeneration in old age may be the price to be paid for youthful evolutionary success. Interestingly enough, there is evidence that the price borne in old age is less onerous in modern societies with a high standard of living than it was in pre-modern times, suggesting that natural selection has indeed weakened as technology has made our environment more benign. Many studies have found that in pre-modern societies, women who had their first children early, and had many of them, died earlier than women who delayed childbearing and had fewer. This pattern disappears as health and standards of living increase.

The long view

In the long run, said the economist John Maynard Keynes, "we are all dead". The same can be said of species. In the long run, the fate of every species, including our own, is to become extinct or replaced. All other species of *Homo*, of which there were until quite recently several (Chapter 5 *African genesis*, and Chapter 14 *The race from Africa*), have already gone the way of the dodo. For us what matters, as Keynes implied, is what happens in the meantime.

Evolution offers us a historical perspective on shorter as well as longer timescales. The fossil record shows that mass extinctions, like the one that extinguished the dinosaurs (other than the ancestors of birds) at the end of the Cretaceous, are caused by catastrophic events that are global in scale. A massive meteor strike, whose impact crater is buried in the Gulf of Mexico, may have caused the extinction of the dinosaurs and many other contemporary species. Before the impact that closed the Cretaceous, our hairy tribe, the mammals, was represented mainly by small, shrew-like animals that were active during the night. Daylight belonged to the reptiles. After the dinosaurs and many other reptiles were destroyed, mammals diversified rapidly (Chapter 8 *A whale of a problem*) and replaced them with species that filled the ecological roles they formerly occupied, from large plant-eaters to carnivores and fliers (bats) and marine species (whales and dolphins). If we owe our distant evolutionary origins to a global catastrophe, will our species be exterminated by similar means and if so, how distant is that prospect?

Astronomers calculate that another meteor impact large enough to threaten humanity's survival is a distinct possibility, although no immediate threats have been detected. Of course, what is different now, compared to the end of the Cretaceous, is that biological evolution has produced a species capable of understanding the origins of some celestial threats, detecting them in advance and possibly using advanced technology to avert some of them too. Unfortunately, the power and intelligence to change the course of planetary history make a two-edged sword.

Evolution in the Anthropocene

An extraterrestrial astronomer, from some inhabited planet circling another star and gazing at our solar system as it was 200,000 years ago, would have had little doubt that there was life on Earth. That distant astronomer, too far away to see life itself but equipped to look for clues in the composition of Earth's atmosphere, would spot that it was unusually rich in oxygen, a sure tell-tale of life (Chapter 4 *First life*). Anatomically modern humans were present in Africa at that time (Chapter 5 *African genesis*), but their numbers would have been small and their impact upon their environment too slight to notice. The signature of life to be read in our atmosphere at that time was penned by photosynthesis, not people. Fast-forward 199,800 years to 1800 AD and the human population had reached about a billion. Impacts on the ground due to forest clearance and agriculture were widespread, but the composition of the atmosphere did not yet register a response to human activities. All that was about to change.

The industrial revolution that began about 200 years ago required power and that need was fed by fossil fuel. In the 19th century it was coal; increasingly in the 20th century it was oil and natural gas. The fossil fuels that power our transport, industrial processes and agriculture have now raised the concentration of carbon dioxide in the atmosphere by more than 36% over the pre-industrial level, and the concentration of this greenhouse gas is rising annually. The fourth report of the Intergovernmental Panel on Climate Change, published in 2007, left little room for doubt: this change in the

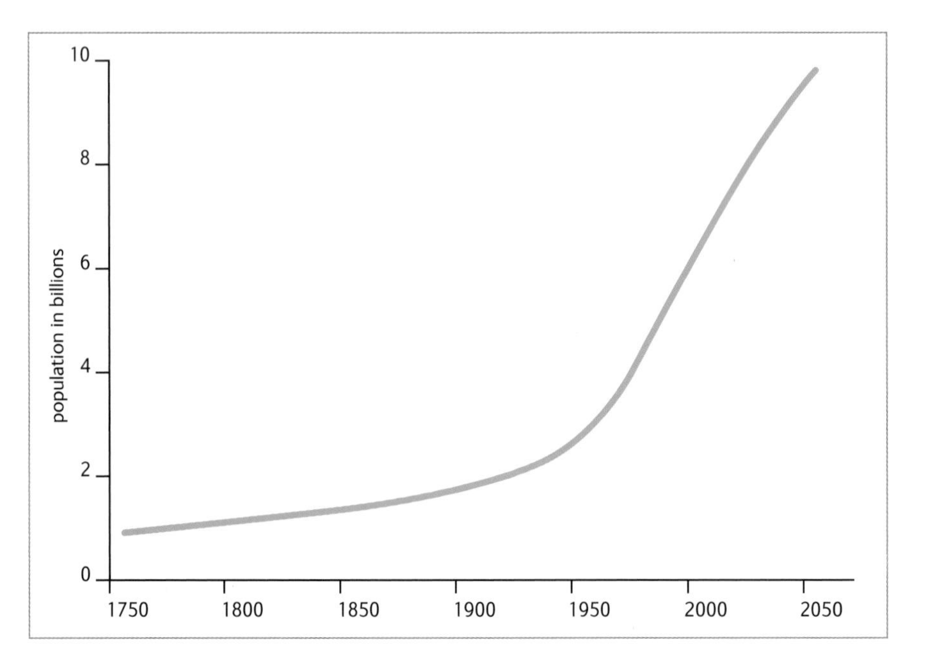

RIGHT The size of the human population has increased dramatically in the last two centuries and is expected to reach 9 or 10 billion in the next 50 years.

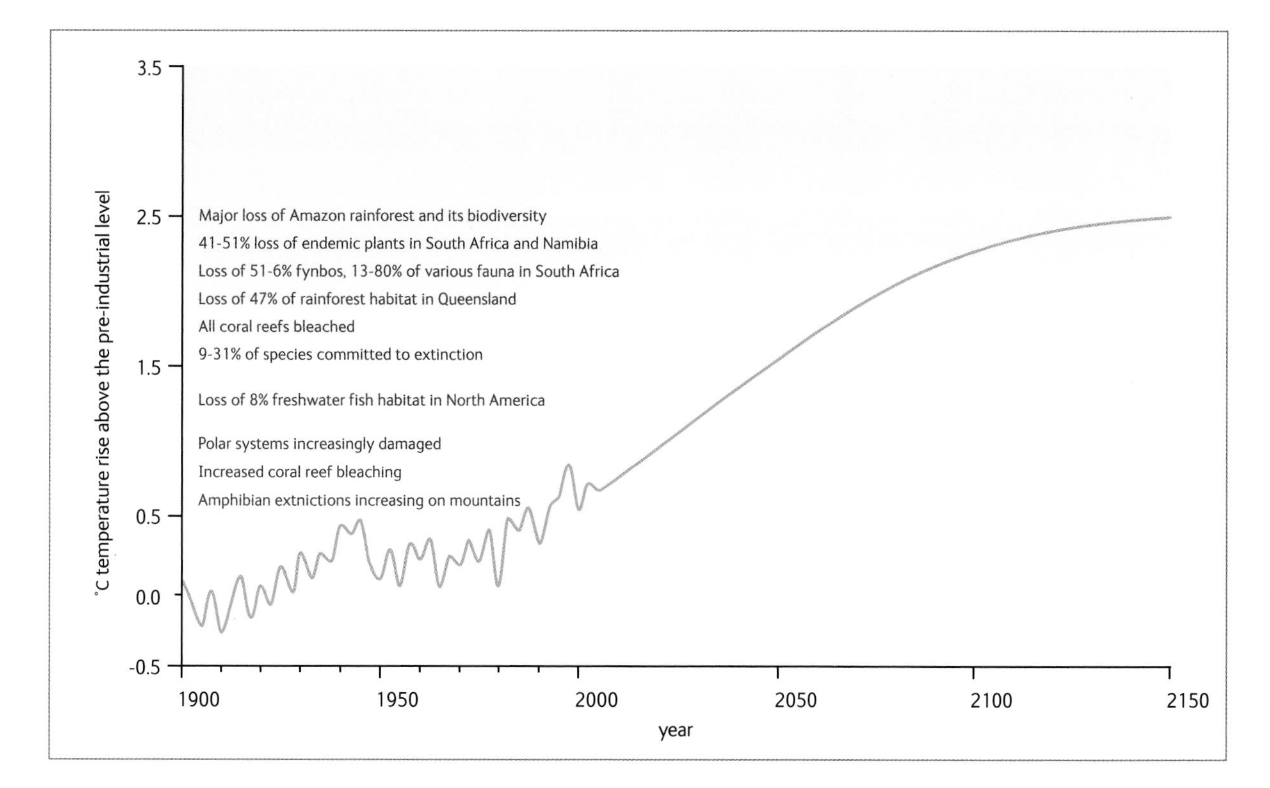

ABOVE Increase in mean global temperature above the pre-industrial level since 1900, and one of the more conservative scenarios for its change over the next century and a half. Some of the consequences of global warming for life on Earth that have been predicted by the International Panel on Climate Change in their 2007 report are shown.

atmosphere is warming the climate throughout the globe, especially at high latitudes. The influence of our species upon the planet and its atmosphere is now so extensive and global in scale that, according to the Nobel Prize-winning atmospheric scientist Paul Crutzen, we have entered a new geological epoch of our very own: welcome to the Anthropocene.

Geologists recognize five episodes when the fossil record shows that the diversity of life on Earth was suddenly and severely depleted. The end of the Permian, 248 million years ago, was the biggest mass extinction when 90–95% of marine species disappeared in a relatively short period. The last mass extinction was at the end of the Cretaceous (65 million years ago) when 85% of animal species went extinct abruptly and were replaced more slowly. Many scientists believe that a sixth mass extinction is occurring now, in the Anthropocene. At the top of the list of most threatened species are our own closest relatives, the great apes.

Chimpanzees, gorillas and orang-utans are imperilled by loss of habitat, human hunting and infection with human diseases (Chapter 15 *Catch me if you can*). It is estimated that, as a species, we utilize 20% of the plant growth that takes place each year on our planet. Perhaps a quarter of all plant species are under threat. Unsustainable fishing threatens all the major commercial fisheries. To cap it all, global warming also threatens many species with extinction. The question we therefore have to ask ourselves about the evolutionary future is not an exercise in science fiction; it is a much more sobering and immediate question about our future reality. What kind of world are we creating, and how can we save as much as possible of the fruits of evolution for our own benefit and for its own sake?

ORIGIN OF SPECIES
DARWIN
DARWIN THE ORIGIN OF SPECIES
ISBN 0 14 040.001 X
Charles Darwin • The Origin of Species
09212
ORIGIN OF SPECIES DARWIN
Darwin
ORIGIN OF SPECIES
DARWIN
A MONOGRAPH OF THE CIRRIPEDIA BALANIDÆ
DARWIN

Sources

CHAPTER 1

Brass, A.L., *et al.* (2008) 'Identification of host proteins required for HIV infection through a functional genomic screen'. *Science*, **319**, 921–926.

Burkhardt, F. and Smith, S. (1987) *The Correspondence of Charles Darwin, Volume 3 1844–1846*. Cambridge University Press, Cambridge.

Centers for Disease Control (1991) Update: 'Transmission of HIV infection during an invasive dental procedure'. *MMWR* **40**(2): 23–27, 33

Ciesielski *et al.* (1992) 'Transmission of human immunodeficiency virus in a dental practice'. *Ann. Intern. Med.* **116**(10): 798–805.

Darwin, C.R. (1859) *On the Origin of Species by Means of Natural Selection*, first edition. John Murray.

de Oliveira, T. *et al.* (2006) 'Molecular epidemiology: HIV-1 and HCV sequences from Libyan outbreak'. *Nature*, **444**, 836–837.

Demuth, J.P. *et al.* (2006) 'The evolution of mammalian gene families'. *PLoS One*, **1**, 1–10.

Human Genome Project – www.ornl.gov/sci/techresources/Human_Genome/home.shtml (accessed 02.05.08)

Mikkelsen, T.S. (2005) 'Initial sequence of the chimpanzee genome and comparison with the human genome'. *Nature*, **437**, 69–87.

Nature web focus – www.nature.com/nature/focus/aidsmedicslibya/index.html (accessed 02.05.08)

Papathanasopoulos, M.A., Hunt, G.M. and Tiemessen, C.T. (2003) 'Evolution and diversity of HIV-1 in Africa – a review'. *Virus Genes*, **26**, 151–163.

Pollard, K.S. *et al.* (2006) 'Forces shaping the fastest evolving regions in the human genome'. *PloS Genetics*, **2**, 1599–1611.

Prabhakar, S. *et al.* (2006) 'Accelerated evolution of conserved non-coding sequences in humans'. *Science*, **314**, 786–786.

CHAPTER 2

Behe, M.J. (2007) *The Edge of Evolution*. Free Press.

Cook, L.M. (2003) 'The rise and fall of the *Carbonaria* form of the peppered moth'. *Quarterly Review of Biology*, **78**, 399–417.

Darwin, C.R. (1859) *On the Origin of Species by Means of Natural Selection*, first edition. John Murray.

Dennett, D.C. (1995) *Darwin's Dangerous Idea*. Penguin.

Lindblad-Toh, K., *et al.* (2005) 'Genome sequence, comparative analysis and haplotype structure of the domestic dog'. *Nature*, **438**, 803–819.

Majerus, M.E.N. (1998) *Melanism: Evolution in Action*. Oxford University Press.

Ostrander, E.A., Galibert, F. and Patterson, D.F. (2000) 'Canine genetics comes of age'. *Trends in Genetics*, **16**, 117–124.

Paley, W. (1802) *Natural Theology*. Reprinted with an introduction and notes by Eddy, M.D. and Knight, D.M. (2006) Oxford University Press.

Parker, H.G., et al (2004) 'Genetic structure of the purebred domestic dog'. *Science*. **304**, 1160–1164.

CHAPTER 3

Darwin, C.R. (1837) 'B' Notebook, viewed online at darwin-online.org.uk (accessed 04.05.08)

Darwin, C.R. (1839) *Narrative of the Surveying Voyages of His Majesty's Ships* Adventure *and* Beagle, *Between the Years 1826 and 1836, Describing their Examination of the Southern Shores of South America, and the* Beagle's *Circumnavigation of the Globe: Volume III – Journal and remarks, 1832–1836*. Henry Colburn.

ST
QR
MN
KL
J
EF

Index

Page numbers in *italic* refer to illustration captions; those in **bold** refer to main subjects of boxed text.

Picture credits

FRONT COVER Image © Jonty Clark

BACK COVER Image © Istockphoto

CHAPTER 1
p.5. © John Reader/Science Photo Library; p.6 © NHMPL; p.7 © Ben Curtis/AP/PA Photos; p.8 © Eye of Science/Science Photo Library; p.9 Mercer Design © The Natural History Museum (redrawn with permission, from the original in Ciesielski et al. 1992. Transmission of human immunodeficiency virus in a dental practice. *Ann. Intern. Med.* 116(10): 798–805). The American College of Physicians are not responsible for the accuracy of the translation; p.10 Mercer Design © The Natural History Museum (redrawn with permission, from the original in Sharp et al, 2000, *Biochemical Society Transactions*, 28, 275-282); p.11 © NHMPL; p.12 top © Richard Doyle/Cartoonstock.com; p.12 bottom © Science Photo Library; p.13 top © NHMPL; p.13 bottom reproduced with permission from John van Wyhe ed., *The Complete Work of Charles Darwin Online* (http://darwin-online.org.uk/); p.15 Mercer Design © The Natural History Museum (redrawn from an original illustration in *Nature*, 1 September 2005, vol. 437, issue no. 7055, pp.18–19.

CHAPTER 2
p.16 © Istockphoto; p.17 © Jonathan Silvertown; p.18 © The Open University; p.19 © NHMPL; pp.22–25 © NHMPL.

CHAPTER 3
p.26 NHMPL; pp.27–28 © Istockphoto; p.30 Mercer Design © The Natural History Museum; p.31 top © NHMPL; p.31 bottom © David Tipling/NHMPL; p.32 © NHMPL; p.33 Mercer Design © The Natural History Museum; p.34 Mercer Design © The Natural History Museum (redrawn from figure 1 in *Palaeontology*, Neraudeau, D., Viriot, L., Chaline, J., Laurin, B. and van Kolfschoten, T., Discontinuity in the Plio-Pleistocene Eurasian water vole lineage, p. 77-85, vol.38, 1995. Reproduced with permission of Blackwell Publishing Ltd.; p. 35 Reproduced by kind permission of the Syndics of Cambridge University Library; p.37 Mercer Design © The Natural History Museum; p.38 © Istockphoto; p.39 Mercer Design © The Natural History Museum.

CHAPTER 4
pp.40–41 © Istockphoto; p. 42 top Mercer Design © The Natural History Museum; p.42 bottom © Fotolibra; p.43 top © Pasieka/Science Photo Library; p.43 bottom © Istockphoto; p.44 © Richard J. Wainscoat, Peter Arnold Inc./Science Photo Library; p.45 top © Take 27 Ltd; p.45 bottom © Mike Eaton; p.47 top © NHMPL; p.47 bottom © Sinclair Stammers/Science Photo Library; p.48 top Mercer Design © The Natural History Museum; pp.48–49 © Istockphoto; p.49 top © Peter Batson, DeepSeaPhotography.com; p.50 top and bottom © ESA: p.51 © NASA.

CHAPTER 5
pp.52 © NHMPL; p.53 Reproduced with permission from The Open University; p.54 © NHMPL; p.55 © Istockphoto; p.56 top Mercer Design © The Natural History Museum; p.56 bottom and p.57 © David Robinson; p.58 © NHMPL; p.59 top and bottom left © Istockphoto; p.59 right © NHMPL; p.60 Mercer Design © The Natural History Museum; p.61 © MPFT.

CHAPTER 6
p.62 © NHMPL; p.63 © Istockphoto; p.64 Mercer Design © The Natural History Museum; p.65 © Istockphoto; p.66 © NHMPL; pp.67–69 © Istockphoto.

CHAPTER 7
p.70 © NHMPL; p.71 © Jean Claude Revy – ISM/PhototakeUSA.com; p.72 © Jim Hannay;

p.73 © Tom McHugh/Science Photo Library; pp.74–75 © Rudie Kuiter; p.76 © NHMPL; p.77 © Peter Scoones/Science Photo Library; p.78 top © Richard Hammond; p.78 bottom © Istockphoto; p.79 top © Ted Daeschler/ Academy of Natural Sciences/VIREO; p.80 bottom Mercer Design © The Natural History Museum.

CHAPTER 8
p.82 © Brandon Cole/Naturepl.com; p.83 © Fotolibra; pp.84–86 top © Istockphoto; p. 86 bottom © Tony Heald/Naturepl.com; p.87 © Istockphoto; p.88 © NHMPL; p.90 top © Istockphoto; p.90 bottom © Martyn L Gorman The Aberdeen University Zoology Museum; p.91 all © NHMPL; p.92 top © NHMPL; p.92 bottom © John Sibbick/NHMPL; p.93 Mercer Design © The Natural History Museum.

CHAPTER 9
p.94 © NHMPL; p.95 left © Gregory Dimijian/Science Photo Library; p.95 right © Caroline M. Pond; p.96 © Istockphoto; p.97 top © Robert C Hermes/Science Photo Library; p.97 bottom © John Sibbick/NHMPL; p.98 top © Institute of Vertebrate Paleontology and Paleoanthropology, Chinese Academy of Sciences; p.98 bottom © Mick Ellison; p.99 © NHMPL; p.100 © The Open University; p.101 Mercer Design © The Natural History Museum; p.103 © The Open University.

CHAPTER 10
p.104 © Robert Mileham; p.107 Mercer Design © The Natural History Museum; pp.108–9 © Istockphoto; p.110 © Pam Eveleigh/Primula World; p.111 © Robert Cameron; p.112 top Lisa Wilson © The Natural History Museum, London (redrawn, with permission, from the original in the *Annual Review of Ecology and Systematics*, vol.8 © 1977 by Annual Reviews www.annualreviews.org; p.112 bottom Lisa Wilson © The Natural History Museum. Redrawn from the original, with permission, in Cain, A. J. and P. M. Sheppard (1954). "Natural Selection in Cepaea." *Genetics* 39(1): 89-116; p.113 © Dave Watts/Naturepl.com.

CHAPTER 11
p.114 © Phil Savoie/Naturepl.com; p.115 © Mules and More; p.116 top Lisa Wilson © The Natural History Museum; p.116 bottom © NHMPL; p.117 left © Barry Mansell/ Naturepl.com; p.117 middle © Eric Secker; p.117 right © Ron Austing; p.118 © Istockphoto; p.119 © Kevin Kaneshiro; p.120 top © Jim Lindsey; p.120 bottom © Jurgen Freund/Naturepl.com; p.121 © University of Amsterdam Library; p.123 Mercer Design © The Natural History Museum.

CHAPTER 12
p.124 © NHMPL; p.126 © David Robinson; p.127 top © David Robinson; p.127 bottom © NHMPL; p.128 Lisa Wilson © The Natural History Museum (redrawn from the original in G. Estes et al., *Darwin in the Galápagos; his footsteps through the archipelago*, pp.343–368, 2000, Notes and records of the Royal Society); p.129 top © David Robinson; p.129 bottom © Peter Grant; p.130 Mercer Design © The Natural History Museum (redrawn from the original in Peter R. Grant, *Ecology and Evolution of Darwin's Finches*, 1986, Princeton University Press); p.131 © DK Images; p.132 Mercer Design © The Natural History Museum (redrawn from the original in *Nature*, 2006, vol.7102, p.565, Arhat Abzhanov et al., The calmodulin pathway and evolution of elongated beak morphology in Darwin's finches); p.133 Lisa Wilson © The Natural History Museum; p.134 Mercer Design © The Natural History Museum (redrawn from the original in Peter R. Grant, *Ecology and Evolution of Darwin's Finches*, 1986, Princeton University Press).

CHAPTER 13
p.136 © NHMPL; p.137 top © NHMPL; p.137 bottom © Dr. Jeremy Burgess/Science Photo Library; p.138 © London School of Hygiene and Tropical Medicine/Science Photo Library; p.139 top © Dr. Morley Read/Science Photo Library; p.139 bottom left © Françoise Gantet (http://www.creaweb.fr/bv); p.139 bottom right © NHMPL; pp.140–142 top © Istockphoto; p.142 left © Jonathan Silvertown; p.143 Mercer Design © The Natural History Museum (redrawn from the original in Judd, W. S., C. Campbell, et al., 2008, *Plant Systematics: A Phylogenetic Approach*, 3rd Edition, Sinauer Associates); pp.144–145 © Istockphoto; p.146 top and bottom © NHMPL; p.147 © H. D. Bradshaw Jr & Douglas W. Schemske; p.148 © NHMPL.

CHAPTER 14
p.150 © NHMPL; p.151 Human Origins Program/Smithsonian Institution; p.152 top and bottom © David Robinson; p.153 top © Debbie Maizels, adapted from *Biology: The Dynamics of Life*, © 2004 The McGraw-Hill Companies; p.153 bottom © NHMPL; p.155 Lisa Wilson © The Natural History Museum; p.157 Mercer Design © The Natural History Museum.

CHAPTER 15
p.160 © Nigel Cattlin/FLPA; p.161 © RHS; p.162 © Steve Gschmeissner/Science Photo Library; p.163 © Kwangshin Kim/Science Photo Library; p.164 © Eye of Science/Science Photo Library; p.165 © Istockphoto; p.166 © Dr. Tony Brain/ Science Photo Library; pp.167–169 © NHMPL; p.170 top © Dr, Chris Hale/Science Photo Library; p.170 bottom Mercer Design © The Natural History Museum.

CHAPTER 16
pp.172–174 © Istockphoto; p.176 © APPI/ Semitour Perrigord – Lascaux II, fac-simile de la grotte de Lascaux (Montignac - Dordogne/ Perigord) www.semitour.com; p.177 top © Copyright 2008 National Academy of Sciences, U.S.A.; p.177 bottom © Istockphoto; p.178–9 © NHMPL; p.180–181 © Istockphoto.

CHAPTER 17
pp.182, 184, 185 bottom © Istockphoto; p.186 © Getty Images; p.189 © Istockphoto; p.190 Mercer Design © The Natural History Museum redrawn from the original in Bateson, M., D. Nettle & G. Roberts, (2006). Cues of being watched enhance cooperation in a real-world setting. *Biology Letters* 3: 412–4; p.191 © Istockphoto.

CHAPTER 18
p.192 Reproduced by kind permission of the Syndics of Cambridge University; p.194 © Sheila Terry/Science Photo Library; p.197 © Istockphoto; p.199 © AP/PA Photos.

CHAPTER 19
pp.200–202 © Istockphoto; p.203 top © Tim Vernon, LTH NHS Trust/Science Photo Library; p.203 bottom © Istockphoto; p.204 left © English Heritage Photo Library, by kind permission of Darwin Heirlooms Trust; p.204 right and 205 © NHMPL; p.206 top © University of Sussex; p.206 bottom right© johnhegley.co.uk; pp.208–209 Mercer Design © The Natural History Museum; p.210 © NHMPL.

(NHMPL, The Natural History Museum Picture Library; MPFT, Mission Palaeoanthropologique Franco-Tchadienne)

Every effort has been made to contact and accurately credit all copyright holders.
If we have been unsuccessful, we apologise and welcome correction for future editions and reprints.

About the authors

CAROLINE POND is Professor of Comparative Anatomy in the Department of Biological Sciences, at the Open University. Richard Dawkins described her as "the zoologists' zoologist".

JONATHAN SILVERTOWN is Professor of Ecology in the Department of Life Sciences at the Open University, and is internationally-known for his research on the evolution and ecology of plants.

DAVID ROBINSON is a zoologist and evolutionary biologist in the Department of Life Sciences at the Open University.

PETER SKELTON is a palaeobiologist in the Department of Earth and Environmental Sciences at the Open University, and is recognised internationally as an authority on fossil bivalves and their evolution.

DANIEL NETTLE worked at the Open University until 2004, and is currently Reader in the Centre for Behaviour and Evolution at Newcastle University.

MONICA GRADY is Professor in Planetary Sciences at the Open University, and one of the world's leading meteorite experts.

GARY SLAPPER is Professor of Law at the Open University, and has a long-standing interest in legal battles over the teaching of Darwinism in the USA.